A-level Study Guide

Geography

Chris Burnett

David Burtenshaw

Nick Foskett

Rosalind Foskett

Garrett Nagle

Lorraine Wadsworth

Emma West

Revision Express

Acknowledgements

Figure 49A – with kind permission of the Dundee Satellite Receiving Station, Dundee University, UK.

Series Consultants: Geoff Black and Stuart Wall

Project Manager: Gillian Ragsdale

Pearson Education Limited

Edinburgh Gate, Harlow

Essex CM20 2JE, England

and Associated Companies throughout the world

First published 2001

10 9 8 7 6 5 4 3 2 1

09 08 07 06 05

British Library Cataloguing in Publication Data

A catalogue record for this title is available from the British Library.

ISBN 1-405-82119-1

Set by 35 in Univers, Cheltenham

Printed by Ashford Colour Press, Gosport, Hants

Geomorphology and Hazards

Geomorphology is the study of the form of the earth. Studies at A-level concentrate on the evolution of landforms and the processes of erosion, deposition and weathering that have formed them. Having first looked at the structure of the earth's crust and the way that it is changing, a short section covers the main processes of weathering and erosion that are to be found acting upon the crust. The approach adopted is to investigate the landforms and processes that are responsible for their creation in a series of different environmental contexts. Hazards may be caused by natural or quasi-natural processes or human actions. In this section the emphasis is upon natural and quasi-hazards as this is the way that most examination syllabuses view the topic.

Exam themes

In this chapter the natural processes and landforms have been dealt with separately from the human management strategies, but the examination boards often amalgamate the two elements as different parts of the same question. Each examination board varies in the way that the questions are set about these themes, but essay questions, data response questions and structured questions are all possibilities. With all questions it is vital that you can explain how the processes that have caused the landform or hazard operate, as well as to know real-world examples of places where these things have actually taken place.

Topic checklist

○ AS   ● A2

	EDEXCEL A	EDEXCEL B	OCR A	OCR B	AQA A	AQA B	WJEC
Plate tectonics	○	●	○	●	○	●	○
Earthquakes and volcanoes	○	●	○	●	○	●	○
Weathering and mass wasting	○	○	○●	○	●		
Glacial systems and processes	●		●	●	●	○	●
Glacial landforms	●		●	●	●	○	
Periglacial processes and landforms	●		●	●	●	●	
Coastal processes and landforms	○	○	●	○	●	○	●
Hydrological cycle and drainage basin systems	○	○	○●	○	○	○	○
River channel and basin processes	○	○	○●	○	○	○	○
Fluvial landforms	○	○	○●	○	○	○	○
Managing natural environments	○	○	●	○	○●		○●
Hazard classification and perception	●	●	●	●		●	
Managing hazards	●	●	●	●		●	
Tectonic, geomorphological and atmospheric hazards	○●	●	●	●	●	●	○●

Plate tectonics

The earth's crust is formed out of a number of blocks or **tectonic plates** that float on a part of the mantle called the **asthenosphere**. Plate tectonics is the study of the movement of these plates and the landforms, submarine features and hazards that result from these movements.

Plate movement

You need to know the evidence that the experts used to prove that the plates move and also the mechanics of how they actually move.

Evidence for plate movement

→ fit of continental coastlines
→ matching of geological structures across oceans
→ matching of climatically controlled rock formations across oceans
→ matching of species across oceans
→ **palaeomagnetism**
→ **sea floor spreading**
→ **mid-ocean ridges**

How plates move

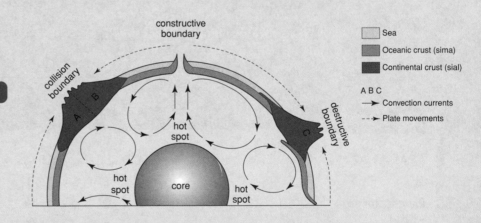

Types of plate boundary

There are three types of plate boundary and the movements of the plates at each of them produce different landforms.

Constructive plate boundaries

At this type of boundary the two plates are moving apart from each other and new crust is being created. As the plates are pulled apart by the forces of the **convection currents** in the mantle below, huge rift valleys usually form on the sea floor. Magma rises to fill any gap between the plates and a **submarine volcano** is produced. If this process continues for many years the height of the volcano may rise above sea level to create an island such as Surtsey in Iceland, which is on the Mid-Atlantic Ocean Ridge.

Destructive plate boundaries

There are three types of destructive boundary where two plates are moving towards each other or converging, usually causing one of them to be subducted into the asthenosphere.

1 Oceanic plate meets continental plate. The oceanic plate subducts under the continental plate and a deep ocean trench forms. Sediments in the area are folded and faulted into a range of fold mountains along the edge of the continental plate. Some 80% of the world's volcanoes are located in these areas.

2 Two oceanic plates converge. When two oceanic plates converge the result is a deep ocean trench and a line of volcanic islands called an island arc.

3 Two continental plates converge. These are rare. Two continental plates are forced together and as they are of similar density and much lighter than the asthenosphere no subduction takes place. The collision forces the sediments between the two plates to be forced upwards into some of the largest mountain ranges in the world.

Conservative plate boundaries

At this type of boundary the plates are being dragged past each other and no crustal material is being created or destroyed. Where this type of movement is taking place there are often a series of **transform faults**. These are at right angles to the main boundary and earthquakes are a major hazard in these locations as much as along the main boundary.

The crust ●●●

There are two types of crust making up the outer layer of the earth.

→ Oceanic crust (sima) is almost continuous around the earth, is relatively dense (3.0–3.3 gm per cc), consists mainly of basaltic rocks and is less than 250 million years old. This forms only a thin layer between 6 and 10 km thick.

→ Continental crust (sial) only occurs where there are continental land masses (on average it is between 35 and 70 km thick), is lighter, only about 2.7 gm per cc, consists mainly of granitic rocks and is much older – in some places more than 3500 million years old. The oldest, most stable and most eroded areas of this type of crust are called **cratons or shield areas**.

Action point

Draw and label a diagram of each type of destructive plate boundary to explain the processes taking place.

Checkpoint 3

Can you accurately locate an example of each of these types of plate boundary?

Watch out!

Often both plates are moving in the same direction at one of these plate boundaries, but at a different speed. This gives the impression that they are actually moving in opposite directions.

Exam questions answers: page 32

1 Using the theories of plate tectonics, explain the major relief features of one continent that you have studied. (20 mins)

2 There are many pieces of evidence that scientists have used to prove the theories of plate tectonics. Briefly describe the most important of these.

Examiner's secrets

When answering questions like these it often saves time if you include well-labelled diagrams.

Earthquakes and volcanoes

Vulcanicity refers to the forcing of solid, liquid or gaseous material into or onto the surface of the earth's crust. These processes are referred to as **intrusive** when the materials are forced into the crust and **extrusive** when the material is forced onto the surface.

Intrusive landforms

Far more magma is intruded into the crust than ever reaches the surface. It cools and solidifies within the crust and after many centuries of **denudation** the various forms of intrusion are exposed as landforms. These vary in size and shape, often depending on the rock structures into which the intrusions take place, and include **batholiths**, **laccoliths**, **phacoliths**, **dykes** and **sills**. As the hot material is injected into the crust from below, the existing solid rock experiences changes due to the heat and/or pressure exerted upon it. The rocks that are changed in this way are referred to as the **metamorphic aureole**.

Extrusive landforms

Major landforms

Volcanoes are the most common landform that result from extrusive volcanic activity and they may be classified in a number of ways:

→ the characteristics of the extruded material (acidic or basic)
→ the nature of the opening or vent through which the lava has emerged (fissure or vent)
→ the frequency of the eruption at that location (regular or infrequent)
→ the degree of violence of the eruption (explosive or gentle).

Volcanoes may occur in a variety of different shapes and sizes depending upon the factors above.

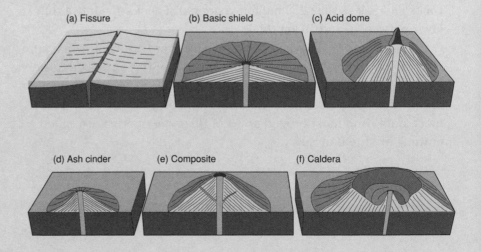

(a) Fissure   (b) Basic shield   (c) Acid dome

(d) Ash cinder   (e) Composite   (f) Caldera

Examiner's secrets

Examiners are really impressed if you can name real examples of all landforms such as these.

Test yourself

Make sure that you can explain which sedimentary rocks are changed into which metamorphic rocks by this process.

Action point

Make copies of each of the different volcanoes in the diagram and add as many labels as you can to explain their characteristic shapes.

Where are volcanoes found?

Volcanoes are found in five different types of location:

→ at spreading mid-ocean ridges
→ along continental rift valleys
→ in island arcs at destructive plate margins
→ in fold mountain ranges
→ in isolated locations known as hot spots.

Minor landforms

Minor landforms such as **mud volcanoes**, **solfataras**, **geysers** and **fumaroles** are formed as a result of extrusive volcanic activity.

Earthquakes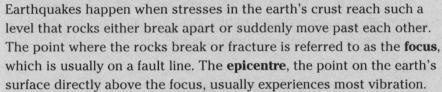

Earthquakes happen when stresses in the earth's crust reach such a level that rocks either break apart or suddenly move past each other. The point where the rocks break or fracture is referred to as the **focus**, which is usually on a fault line. The **epicentre**, the point on the earth's surface directly above the focus, usually experiences most vibration.

Energy is released from the focus in a series of different types of waves, the study of which has enabled scientists to develop a greater understanding of the internal structure of the earth. Most earthquakes occur at plate boundaries, the majority of the largest take place at subduction and collision zones but they also happen at mid-ocean ridges and along conservative plate boundaries such as the San Andreas Fault. As many earthquakes occur under the sea, **tsunamis** are an associated hazard in coastal areas. Minor earthquakes also occur:

→ near volcanic eruptions
→ along old fault lines, often due to **isostatic readjustment**
→ when stresses in the crust have been released by filling valleys with very heavy loads of water when building huge reservoirs.

Measuring earthquakes

There are two scales used to measure earthquakes. The **Richter scale** is a logarithmic scale, which goes up to 10, of the amount of energy released. The **Mercalli scale** is a measure of the amount of ground movement that occurs.

Checkpoint 1

Explain why volcanoes are found at each of the locations listed.

Test yourself

You should be able to describe and explain the formation of each of these minor extrusive features.

Action point

Draw your own diagram to show the relationship between the focus, epicentre and magnitude of vibrations of the crust associated with an earthquake.

Checkpoint 2

What are the three different types of earthquake waves?

The jargon

A *logarithmic scale* is divided into cycles, each being ten times greater than the one before. This means that an earthquake of Richter scale 6 is ten times greater than one of Richter scale 5.

Examiner's secrets

Your answers will be much improved if you can give appropriate facts and figures (such as the magnitude on the Richter scale) about real examples of earthquakes that you have studied.

Exam questions   answers: page 32

1   Discuss the opinion that vulcanicity can be said to be both hazardous and advantageous to people. (20 mins)

2   Why do earthquakes associated with the Mid-Atlantic Ridge usually have a shallow focus whereas those in Japan have a deep focus? (10 mins)

3   Most volcanoes are located at, or close to, plate boundaries, but the islands of Hawaii are volcanoes in the centre of the Pacific plate. Explain this apparent anomaly. (10 mins)

Weathering and mass wasting

The landforms of the earth's surface have been, and are still being, shaped by the processes of **denudation**. First comes the breakdown of the solid material of the earth's crust *in situ*, which is called weathering. The loose debris is then removed by various processes of erosion depending upon the climate and rock type in the location.

Weathering

Weathering is the disintegration and decomposition of rocks *in situ* by the actions of weather, plants and animals. There are three categories.

Physical weathering

This is the breakdown of rocks into smaller fragments by mechanical processes without any changes in their chemical composition. Two of the main types involve temperature changes:

→ freeze–thaw
→ insolation weathering or thermal fracturing.

There are also types involving no temperature change:

→ salt crystal growth within the pore spaces of a rock
→ pressure release or dilation.

These processes widen the joints and cracks in the rocks until pieces break off. The broken fragments of a variety of shapes and sizes are called the **regolith** and this often covers the surface of the bedrock.

Chemical weathering

This is the decomposition or rotting of rocks due to chemical changes:

→ oxidation
→ reduction
→ carbonation
→ solution
→ hydrolysis
→ hydration.

Biological weathering

This is the breakdown of rocks by the action of plants and animals. It may be physical (mechanical), such as the action of tree roots breaking rocks apart, or chemical, such as when organic acids are released from decaying organic matter and the rocks beneath begin to decompose.

Factors affecting the rate of weathering

→ Climate – temperature and precipitation totals and variability.
→ Rock type – some rocks are more vulnerable to erosion than others.
→ Human activities including pollution – human pollution causes weathering rates to increase, particularly through acidification.
→ Rock structure – degree of jointing and angle of dip.
→ Relief – how high or exposed the land is.

The jargon

In situ is a Latin phrase meaning 'in place' and here it is used to mean that there is no movement involved.

Watch out!

Soil is not regolith, but a thinner layer that also involves biochemical processes.

Checkpoint 1

Can you give an example of a type of rock that is easily weathered by chemical erosion and explain the processes involved?

Mass movement or mass wasting ●●●

This is the movement of material downslope due to gravity, which is often almost imperceptibly slow. Water is usually present in the regolith and lubricates the particles of weathered material. Classifications of flows may be by speed, i.e. fast to slow, or by type of movement, i.e. flow, slide or heave. The most accepted classification of mass movements is that of Carson and Kirkby (1972) shown below.

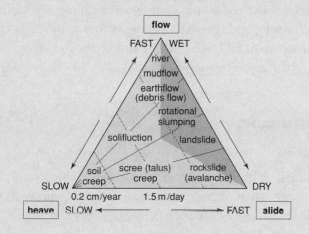

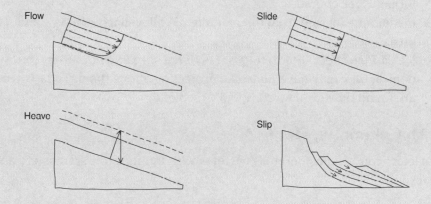

Factors influencing mass movement

→ Internal strength of regolith
→ Volume of loose material present on the slope
→ Slope angle
→ Water content of material/rainfall amounts
→ Human activity
→ Earthquakes

Exam questions   answers: page 32

1   Explain the climatic conditions under which mechanical and chemical weathering are most effective. (10 mins)

2   Explain why mass movements are difficult to classify. (10 mins)

3   Describe the landforms that result from the weathering of either granite or limestone and the processes involved. (20 mins)

Examiner's tip

It is always an impressive feature of an answer to be able to refer to pieces of research by name and date.

Action point

Construct a table of the types of mass movement shown in the top diagram and give a brief explanation of each. Make sure you think about the ways that each movement is different from others.

Checkpoint 2

With reference to the bottom 4 diagrams, explain the difference between flow, slide, heave and rotational slips.

Examiner's secrets

When describing landforms it is vital that you can use the names of real examples.

Glacial systems and processes

Today glaciers, ice sheets and ice caps cover about 10% of the land surface of the earth whereas during the Quaternary era this figure rose to 30%. Where areas are covered by ice, the land is being shaped by a number of unique processes.

The formation of ice ●●●

Glacial ice is formed as a result of the accumulation of layers of snow above the permanent snow line. Temperature changes lead to freezing followed by thawing and the snow becomes firn or névé. As the increased pressure of additional layers gradually squeezes out the air, the density of the material increases until after 20–40 years it becomes glacial ice.

Checkpoint 1

Why will it probably take several hundred years for this process to take place in Antarctica or Greenland?

Types of glacier ●●●

Glaciers are classified according to their size and shape.

→ Niche – small, in gullies on shaded hillsides.
→ Cirque – larger than niche, in armchair-shaped hollows in mountains; the glacier may spill out to feed valley glaciers.
→ Valley – larger masses of ice, moving downhill, usually following a former river course.
→ Piedmont – formed from the merging of valley glaciers in lowland areas.
→ Ice cap/ice sheet – enormous areas of ice spreading out from central domes; they may have nunataks projecting above them. Today these are found only in Antarctica and Greenland.

The jargon

A *glacier* is a slowly moving mass of ice that originated from an accumulation of snow.

The glacial system ●●●

Glaciers can be considered as systems with inputs, stores, transfers and outputs.

Action point

Make a table of the inputs, stores, transfers and outputs from a glacier system.

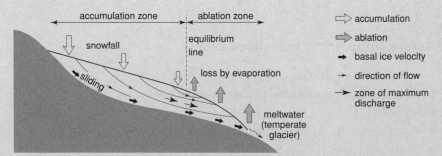

During a year the mass of the glacier will change considerably as shown in the diagram below. This is known as the glacial budget.

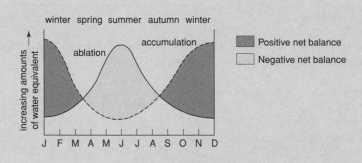

10

Glacial processes

Glaciers are extremely powerful agents of erosion. Erosion is most effective when the ice is moving and contains debris. Thus **warm glaciers** found in the temperate mountain areas are more effective than the **cold glaciers** of polar and subpolar regions.

Ice movement

Ice moves dramatically as **avalanches**, by **basal sliding** along the valley floor (important in warm glaciers) and by **internal plastic flow**. Movement is influenced by friction with the valley floor and sides so fastest movement is in the middle of the valley at the surface of the ice.

The processes of glacial erosion

→ Plucking – tearing away previously weathered pieces of bedrock.
→ Abrasion – polishing, scratching and gouging of the bedrock.
→ Frost shattering – breaking off pieces of bedrock by the alternating of freezing and thawing of water within joints in the rock.
→ Rotational movement – increased pressure resulting from pivoting of ice in a hollow.
→ Dilation/pressure release – upward expansion in rocks beneath the ice due to reduction in the overlying mass as rocks are eroded causes joints parallel to the surface.
→ Meltwater erosion – normal fluvial erosion below and beyond the ice.

The processes of glacial transportation

Anything from fine sediments to large boulders can be carried.

→ Supraglacially – on the surface of the glacier.
→ Englacially – carried within the glacier or ice sheet.
→ Subglacially – moved along at the base of the glacier.

The processes of glacial deposition

Drift is the name used to describe all material that is deposited under glacial conditions. There are two types:

→ **Till** is all material deposited directly from the ice. It is unsorted. It includes **boulder clay**, **erratics**, **moraines** and **drumlins**.
→ **Fluvioglacial deposits** are laid down by meltwater either within or beyond the ice and are sorted materials, often arranged in layers. These include **outwash sands and gravels**, **varves**, **kames** and **kame terraces**, **eskers**, **kettles** and **braided streams**.

Exam questions   answers: page 33

1   Describe how and explain why the mass of a glacier changes during the course of a year. (15 mins)

2   Explain how global warming will affect the mass balance of an alpine glacier. (10 mins)

3   Describe the characteristics and typical location of a drumlin. (10 mins)

Checkpoint 2

Can you explain why a warm glacier erodes more than a cold glacier?

Checkpoint 3

You should be able to describe how each of these three methods of ice movement takes place.

Action point

Can you explain why it is that these processes are likely to be most effective when temperatures frequently alternate above and below 0°C, rocks are well jointed, gradients are steep and where ice depths are great?

The jargon

Erosion is wearing away. *Transportation* is material being carried along. *Deposition* is material being deposited.

Examiner's secrets

As with all A-level questions in geography the examiners are always going to credit your answer more highly if it includes appropriate examples.

Glacial landforms

Some of the most dramatic mountain scenery in the world is the result of the glacial processes described on the last two pages. Glaciers have also been responsible for shaping many of the lowland areas of the world and glaciated landscapes have had an important impact on human activity.

Landforms of glacial erosion ●●●

Many of the features of glaciated erosion are to be found in highland areas. The effect of weathering and erosion since the last ice age has altered the appearance of many of these landforms but this effect has been less significant where the last ice has disappeared relatively recently.

Nivation hollows are small hollows, usually containing a patch of snow. These become enlarged by the process of nivation. In locations where the climate allows the snow to remain throughout the summer these may develop into cirques.

In the highest areas there are **pyramidal peaks**, **cirques**, and **arêtes**. The most impressive landform is the **glacial trough** with its associated **hanging valleys**, **truncated spurs** and **rock steps**. When these features are dammed, **ribbon lakes** form and if they are flooded by a rising sea level a **fjord** results. Within the large glacial troughs smaller features such as **roche moutonnée**, **crag and tail** and **striations** are to be found. During the main phase of glaciation the effects of erosion within the upland areas can be so immense that watersheds are breached by glaciers, with the result that original drainage patterns are altered and rivers may end up flowing in a different direction after the period of glaciation. This is **drainage diversion** or **river capture**.

The formation of a cirque

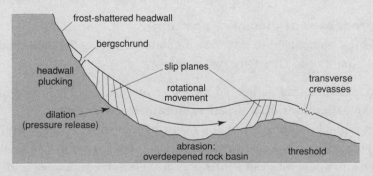

frost-shattered headwall

bergschrund

headwall plucking

slip planes

rotational movement

transverse crevasses

dilation (pressure release)

abrasion: overdeepened rock basin

threshold

How glacial landforms can influence human activity

→ Provide routeways for communication in highland areas
→ Tourism income from visitors to spectacular scenery
→ Mineral extraction/quarrying
→ Dam sites for HEP
→ Frequently used for reservoir construction and water supply
→ Transhumance agriculture (now in decline in most areas)

Landforms of glacial deposition

Till deposits – unsorted material (unstratified)

→ Boulder clay
→ Lodgement till
→ Ablation till
→ Erratics
→ Moraine – lateral, terminal, push, medial
→ Drumlins

Glacifluvial deposits – sorted material (stratified)

Beyond the snout of the glacier lies an area made up of sands, gravels and clays called an **outwash plain** or **sandur**, formed by deposition of material by meltwater streams during the summer or as the glacier melts. Other features include those shown in the diagram below and:

→ **varves** – annually deposited alternating layers of sediments
→ **kettles** – hollows or depressions in glacifluvial deposits resulting from the melting of a block of ice that was trapped in the deposits
→ **braided streams** – choked due to seasonal discharge changes
→ **kames** and **kame teraces**
→ **eskers**.

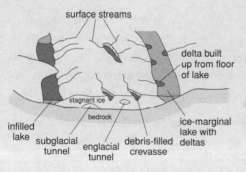

(a) Glacial landscape

(b) Postglacial landscape

Links to human activity

→ Mineral extraction for building materials and subsequent landscape reclamation
→ Types of agriculture linked to variety of soil types
→ Location of settlements linked to landforms
→ Mountain wilderness areas and national parks – sustainability

Exam questions

answers: page 33

1   Describe the characteristics of a roche moutonnée and explain the processes that have led to its formation. (10 mins)

2   Explain how Fluvioglacial deposits are different from glacial deposits. (5 mins)

Checkpoint 2

Describe how and explain why the characteristics of terminal, lateral and push moraines differ.

The jargon

Till deposits are materials dropped by the glacier. *Fluvioglacial deposits* are from water running out of the glacier.

Checkpoint 3

Use the diagram to help you explain the formation of eskers, kames and kame terraces.

The jargon

Sustainability refers to economic growth and social progress that does not damage the prospects of future generations.

Examiner's secrets

Examination questions about glacial features often ask you to explain why the features are located where they are on the landscape.

Periglacial processes and landforms

The term periglacial literally means at the edge of a glacier or ice sheet. However, it is usually used to describe areas that experience very cold climates and a summer thaw. These areas include the tundra areas of the Northern Hemisphere, high mountainous areas and places that were at the edge of the ice sheets during the Pleistocene ice ages.

Permafrost

This is permanently frozen ground. It covers between 20 and 25% of the earth's surface. It develops due to:

→ very cold temperatures (below –5°C) all year round.
→ low precipitation totals so that snow cannot insulate the ground.
→ a limited vegetation cover.

There are several different types of permafrost:

→ continuous
→ discontinuous – islands of permanently frozen ground are separated by less cold areas near the sea, rivers or lakes
→ sporadic – a few isolated areas of frozen ground.

The **active layer** above the permafrost thaws out for a few months and it is here that most periglacial processes take place. It varies in depth from a few centimetres to a few metres. Unfrozen material within or below the permafrost is known as **talik**.

Periglacial processes

Some processes that operate in periglacial environments are unique to these areas whilst others are common in other environments. They are all caused by the alternate freezing and thawing of the ground.

→ **Solifluction** is the slow flow down a slope of material in the active layer that has been saturated or well lubricated by meltwater. The resulting deposit is called **head**.
→ **Wind action** is important in transporting and depositing fine deposits such as glacial and glacifluvial sands. It is most effective in flat areas with very little vegetation cover. Large deposits of this wind-blown sediment are known as **loess** and stretch across Northern Europe and China.
→ **Frost shattering** is most effective where there are frequent freeze–thaw cycles. Water enters cracks or pores in the solid rock and then freezes. Freezing water expands by about 10%, exerting pressure on the rock, eventually causing it to fracture. On steep, exposed rock faces this results in a layer of angular fragments at the base of the face called **scree**. **Blockfields** are large areas of angular rock fragments formed by the effects of frost action, *in situ*, on expanses of flat, exposed rock.
→ **Fluvial action** (erosion, transportation and deposition) is carried out by meltwater in the short summers, but the effects are minor.
→ **Frost heave** and **stone sorting** (see diagram opposite, top).

Checkpoint

Can you explain why the type of permafrost varies from place to place?

Action point

Draw up a table showing each of the main periglacial processes and the landforms that result from their action.

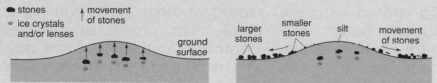

stones forced to the surface by frost heaving – subsequent capillary action of water gives more freezing and enlarges ice lenses

Periglacial landforms

These are some other important periglacial landforms:

→ **Pingos** are relatively large, ice-cored hills that have been domed up by the freezing and consequent expansion of water beneath it. There are two types:
  → open-system caused by the inward freezing of water-saturated sediments
  → closed-system caused by the freezing of water that has moved upwards through a thin permafrost layer.
  The surface of the pingo may fracture, resulting in melting of the subsurface ice and leaving a meltwater-filled hollow.

→ **Ice wedges** are deep, ice-filled, V-shaped cracks in the ground. They are formed as rapid freezing causes the soil to contract, in the shape of polygons. During the following summer the cracks fill with meltwater and other deposits. Each winter as the water refreezes the wedge becomes wider. The pattern of cracks and wedges produces **patterned ground**, **stone stripes** and **stone polygons** (**or circles**).

→ **Thermokarst** is the irregular and hummocky landscape with marshy or lake-filled hollows that results from periglacial conditions. The main cause is the melting of large amounts of ground ice causing surface subsidence. These changes can happen as a result of climatic change or due to human activity.

Test yourself

Can you explain how each of the periglacial landforms described on these pages has been formed?

Action point

Draw a series of labelled diagrams to explain the formation of patterned ground.

Exam question   answer: page 33

The diagram below shows a number of periglacial landforms. Describe the processes that have created the four named landforms. (15 mins)

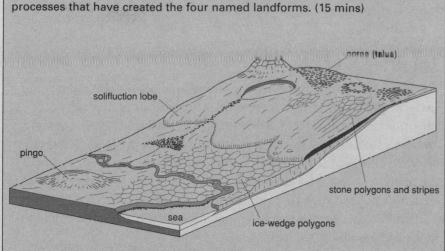

Examiner's tip

If you can draw diagrams as part of your answer it will usually make it much easier to describe landforms or explain processes more clearly.

15

Coastal processes and landforms

Compared with other geomorphological environments, coasts change very rapidly. The processes involved here are very active – complete features such as mud banks and spits may form in little more than a century. Coastal erosion is removing cliffs and beaches at such a rate that local authorities are faced with significant management problems.

Links

Coastal management, page 25.

Coastal processes

Waves are generated by the transfer of energy from the wind blowing across the sea surface. Their size and strength are related to:

→ wind speed
→ fetch (the distance over which they have travelled)
→ wind persistence.

Forced waves are driven ashore by the wind and are of two types:

→ constructive – flatter, longer periodicity (10 or more seconds between waves); these tend to move material up the beach
→ destructive – steeper, more frequent (3 to 6 seconds between waves); material is combed down the beach.

As waves approach the coast their direction is changed by friction with the sea floor to become nearly parallel to the shore. This is **refraction**. Material is transported along the beach by **longshore drift**. The movement of the wave up the beach is **swash** and the return movement down the beach is **backwash**.

Coastal erosion processes

→ Wave quarrying or hydraulic action
→ Abrasion/corrasion
→ Attrition
(also **sub-aerial weathering**, **mass movement** and **human activity**).

The rates of erosion on a coast are influenced by the following:

→ rock lithology, structure and dip
→ local winds and water movement
→ the configuration of the coast leading to **concordant** or **discordant** coastlines.

Landforms of coastal erosion

The two main features produced by wave erosion are **cliffs** and **wave-cut platforms**. The cliff retreats as a result of undercutting and subaerial weathering, leaving behind a platform that is gradually lowered by corrasion. As the platform develops in a seaward direction the wave energy at the cliff face is reduced and consequently so is the rate of cliff erosion. The erosive power of waves is also concentrated on headlands of hard rock that retreat. The waves are able to attack the headland on three sides resulting in **geos**, **caves**, **blowholes**, **arches**, **stacks** and **stumps**.

The jargon

Waves are described using the terms *height*, *velocity*, *length* and *period*. You should be able to use these terms confidently.

Action point

Draw a simple, labelled diagram to show the difference between the strength of the swash and backwash of constructive and destructive waves.

Checkpoint 1

Write one or two sentences to explain each of these processes.

The jargon

The *lithology* of the rock means its relative hardness, permeability and solubility.

Checkpoint 2

Explain the retreat of a headland and the landforms that you would expect to find at various stages in this process.

Landforms of coastal deposition

Material is deposited along the coast when the load-carrying capacity of the sea is reduced, usually as a result of a reduction in wave energy and velocity. Many different landforms result.

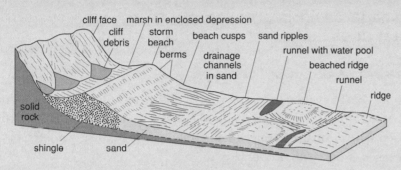

There are a number of other coastal depositional features:

→ spits and cuspate forelands
→ on-shore and off-shore bars and tombolos
→ salt marshes
→ sand dunes.

These features do not rely solely on the erosion of headlands for their supply of sediments. They are also supplied with:

→ material moved landwards from the offshore zone
→ river-borne sediments from inland
→ material from other beaches moved by longshore drift
→ beach nourishment.

Coastal changes

Many coastal landforms are largely the result of changes in the balance between erosion and deposition. This may be the result of a **change in sea level**, which may be either **isostatic** or **eustatic**. When sea level rises, the coastline is said to be **submerged** or **retreating** and lower parts of river valleys may be flooded to create **rias** or **fjords** and cliffs may be subject to renewed wave erosion. When sea level falls, **emerged** or **advancing** coastlines form, with landforms that were created at sea level but are now abandoned above sea level. These include **raised beaches**, **marine** or **raised platforms** and **abandoned cliffs**.

Other changes in the coastal system may be the result of human activities such as land reclamation, spoil tipping, mining, beach control, dredging, beach nourishment and gravel extraction. **Global warming** is another major factor influenced by human activity.

Exam questions
answers: page 33

1   Explain why beach features are transitory parts of the landscape. (5 mins)

2   Describe the process of longshore drift and one landform that is largely created by this process. (10 mins)

3   Use examples to explain the difference between concordant and discordant coastlines. (10 mins)

Don't forget

The features of beaches that are shown here are all transitory because tides, winds and weather conditions are continually changing.

Action point

Draw up a table of each of these features of coastal deposition and the location of an example that you have studied.

The jargon

Isostatic changes are localised changes due to such things as post-glacial uplift of land. *Eustatic* changes are global in extent and affect all oceans equally.

Action point

Can you say whether each of the human influences on the beach environment would tend to cause more erosion or more deposition?

Hydrological cycle and drainage basin systems

The global hydrological cycle is a closed system by which water is transferred between the atmosphere, the oceans and the landmasses. The land surface is divided into a series of drainage basins or catchment areas that supply the river systems with water. The drainage basin is an example of an open system as water does leave and enter this system.

The jargon

A *closed system* is a set of interrelated components within which there are energy transfers but not between the system and its surroundings.

The jargon

A *drainage basin* is an area of land bounded by a watershed, within which all surface and subsurface water will eventually find its way into the same river.

Action point

Draw a systems diagram of a drainage basin showing the inputs, stores, processes (or transfers), and outputs listed in the table.

Checkpoint 1

Explain what will happen to the water table when the water balance is negative, and why it happens.

Action point

List the characteristics of a drainage basin that might influence the soil moisture balance over a year.

Drainage basin systems

Drainage basin systems are a vital part of the hydrological cycle as they return water that arrives on the land surface from the atmosphere to the sea. The drainage basin system contains the following inputs, processes, stores and outputs:

Inputs	Outputs
→ Precipitation → Solar energy	→ River runoff → Evaporation → Transpiration
Stores	**Processes or transfers**
→ Interception → Groundwater → Soil water → Surface storage → Channel storage → Vegetation storage	→ Infiltration → Channel flow → Throughflow → Percolation → Base flow or groundwater flow → Stemflow → Surface runoff or overland flow

The water balance

The water balance is the relationship between the inputs and outputs of a drainage basin. The following equation

$$P = E + R \pm S$$

shows the proportion of the precipitation that enters the system and is lost through evapotranspiration, runoff and changes in the groundwater store. When the balance is positive there is more water entering the system than there is being lost through evapotranspiration and runoff, so groundwater volumes will rise. When it is negative the opposite is true. Periods of surplus, deficit and groundwater recharge can be identified from graphs showing precipitation and evapotranspiration through the year (soil moisture graphs).

To understand these you must be able to explain the following terms:

→ potential evapotranspiration
→ soil moisture recharge, utilisation, surplus and deficit.

The seasonal variations in the water balance influence the following:

→ river regimes
→ springs – whether they are permanent or intermittent
→ the occurrence of ephemeral streams in wet seasons.

Humans alter the water balance by utilising water from the system, depleting it and causing river levels to fall.

Storm hydrographs ●●●

A hydrograph shows changes in river discharge over a period of time. A storm hydrograph shows the way that the discharge at one point along a river varies following a storm event in the catchment area.

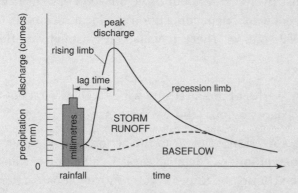

The way a drainage basin responds to a storm event can vary according to:

→ the size of the basin
→ the shape of the basin
→ the drainage density within the basin
→ the porosity and permeability of the soils
→ the geology underlying the basin
→ the steepness of the slopes in the basin
→ the type of vegetation cover in the basin
→ the type of land use, particularly the urban areas within the basin
→ the level of residual moisture in the soil layer
→ the amount, intensity and duration of the rainfall
→ the seasonal weather conditions.

River regimes ●●●

The regime of a river is the way that its discharge varies through a year. A regime may follow a simple pattern with only one peak or it may be more complex and have several peaks. It is very unusual to find a river with a completely natural regime as most rivers are managed or regulated by people for a variety of different purposes. The factors that influence the shape of a storm hydrograph (listed above) may also influence the regime of the river.

Exam questions   answers: page 34

1   Explain how the type, intensity and duration of precipitation affect the relationship between runoff and throughflow. (15 mins)

2   Describe how and explain why the lag time of a river in a catchment area of deciduous woodland would change from summer to winter. (10 mins)

3   Explain how different sizes and shapes of drainage basins may affect the lag time shown on a hydrograph for a place near the mouth of a river. (10 mins)

Action point

Make a list of the features of the storm hydrograph that are labelled on this diagram, and explain the meaning of each feature.

The jargon

Some hydrographs are described as *flashy* if they respond very quickly to storm events.

Action point

All drainage basins are different but you should be able to explain how a variation in each of the variables listed here will affect the shape of the storm hydrograph.

Checkpoint 2

Describe how the building of a dam upstream would affect the regime of a river.

Examiner's secrets

Many examination papers require you to use examples from a drainage basin that you have studied in detail to answer questions on these topics.

River channel and basin processes

Rivers will try to adopt a channel and basin shape that allows them to fulfil their role of transporting water and sediment most efficiently.

River channel processes

Transportation

Rivers transport materials by **traction**, **saltation**, **suspension** and **solution**. Most rivers carry about three-quarters of their load as suspended sediments, depending upon the local rock type, the climate, and the velocity of flow. There may also be seasonal variations.

Erosion

Erosion is the series of processes by which a river wears away its banks and its bed. There are four different processes:

→ **abrasion, corrasion** – similar to the action of sandpaper
→ **corrosion/solution** – a chemical process affecting certain rocks
→ **hydraulic action** – the force of water moving loose material and compressing the air in cracks in the banks and loosening material
→ **attrition** – the breaking down and smoothing of particles as they are rubbed and knocked against each other.

The **critical erosion velocity** is the approximate velocity needed to pick up or **entrain** and transport particles of various sizes.

Deposition

As a river slows down it loses energy and thus its **competence** or **capacity** to transport its load is reduced.

→ Largest particles are deposited first.
→ Finest particles may not be deposited until the river enters the sea.
→ Dissolved load is not deposited, but makes the oceans salty.

Deposition happens when:

→ there is a period of low discharge
→ velocity of flow falls as the river enters a lake or the sea
→ shallow water is encountered, e.g. on the inside of a meander bend
→ the load is suddenly increased, e.g. following a landslide
→ velocity falls outside the channel as a river overflows its banks.

The Hjulström curve

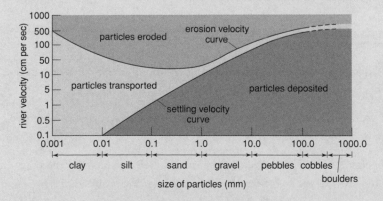

Checkpoint 1

Describe the characteristics of particles carried by each of the four processes of transportation.

Checkpoint 2

Can you explain the meaning of the terms *competence* and *capacity* of a river?

Action point

Choose one of the occasions when deposition happens and explain in detail why it takes place.

The jargon

This diagram shows the relationship between the *velocity of flow* of a river and the size of the particles that it *erodes*, *transports* and *deposits*.

Types of flow

→ Laminar flow – horizontal movement (so rare it is usually discounted).
→ Turbulent flow – consists of a series of horizontal and vertical eddies.

If the velocity of a river is high, there is more energy in the river and so turbulence increases. This causes more erosion as the sediments on the bed are disturbed and moved downstream.

River velocity is affected by:

→ channel shape, which is measured by the hydraulic radius
→ roughness of the bank and bed, which can reduce the velocity of the flow through friction – although there will also be more turbulence so more material will be entrained
→ channel slope becoming less steep as the river reaches its mouth.

Base level and a graded river

The base level is the lowest level to which a river may erode; for most rivers this is sea level. Temporary or local base levels also occur when a river flows into a lake, or where there is a resistant band of rock across the valley. Base levels may be the result of:

→ climatic changes – the effects of glaciation or changes in rainfall pattern
→ tectonic changes – crustal uplift or local volcanic activity.

A positive change is when sea level rises in relation to the land; a negative change is the opposite. If base level changes in any way due to either **eustatic** or **isostatic changes**, the river will adjust its long profile accordingly in an attempt to achieve a **graded profile**. If the base level falls, the **rejuvenation** process starts at the sea and works its way upstream; the point at which the change is occurring is the **knickpoint**. Rejuvenation also leads to the formation of **river terraces** and **entrenched** or **ingrown meanders**.

A river with a graded state will also have its cross-profile and channel roughness in balance with the discharge and load.

Exam questions — answers: page 34

1. Explain what will happen to the relative amounts of deposition and erosion as a result of a positive change in base level. (7 mins)

2. Explain how erosion and deposition will tend to create a graded profile in the river shown below. (10 mins)

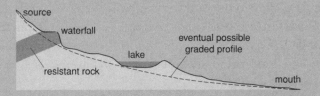

The jargon

The *hydraulic radius* is the measure of the efficiency of a channel and is calculated as the ratio between the area of the cross-section of the river channel and the length of its wetted perimeter.

Action point

Draw labelled diagrams to show where the velocity of flow is greatest in a symmetrical and an asymmetrical channel.

Action point

Find out about Manning's equation, which is used to calculate channel roughness.

Checkpoint 3

When may the base level of a river not be sea level? (One example is the river Jordan, which drains into the Dead Sea.)

The jargon

A *graded profile* is a smooth, concave, long profile of a river. It represents a river in equilibrium, with all its energy being used to move water and sediment and having no free energy to undertake further erosion.

Fluvial landforms

The processes that are operating in the river channel produce a variety of distinctive landforms. These can be classified according to their location within the basin or to the main processes that are responsible for their formation. Neither classification is perfect as many features can be found in different locations within the basin and many features are not the exclusive result of either erosion or deposition. Here the features are described according to their more common location within the drainage basin.

Features of the upper valley

In the upper reaches of the valley the river channel is usually rocky and full of angular boulders. The river channel is therefore very inefficient and there is little energy to pick up and transport material. When the river discharge increases there is more energy available and, as vertical erosion is the dominant process, a **V-shaped valley** with **interlocking spurs** results. As the turbulent water swirls around, pebbles are responsible for scouring out depressions that develop into **potholes**.

The upper sections of a river also contain:

→ **waterfalls and rapids**
→ **gorges**
→ **riffles and pools**
→ **springs**.

Action point

Can you describe and explain the formation of each of these features?

Features of the middle valley

Meanders are found in all sections of the valley and are a result of the processes of erosion, transportation and deposition within the channel. There is some doubt about the exact processes that form them, but the alternate riffles and pools that are found in relatively straight sections of rivers and **helicoidal flow** are thought to be involved.

The jargon

Helicoidal flow is the corkscrew-like movement of water molecules as they move downstream.

Checkpoint 1

Explain how oxbow lakes or cut-offs are formed.

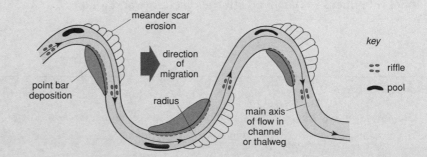

Floodplains are formed beyond the banks of the river by deposition of layers of fine fertile silt at times of flood. As the water rises above **bankfull stage** the hydraulic radius and friction increases resulting in deposition. The floodplain may be widened by lateral erosion of meanders up to the **bluff line**.

Features of the lower valley

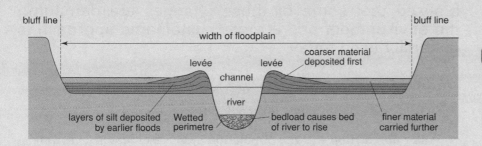

bluff line   width of floodplain   bluff line

coarser material deposited first

levée   levée

channel

river

layers of silt deposited by earlier floods   Wetted perimetre   bedload causes bed of river to rise   finer material carried further

Deltas

As a river nears base level it slows down and deposition becomes the dominant process. The deposited material gradually builds up into a swampy plain known as a **delta**.

The main processes of deposition in this environment are:

➜ bedload dumping
➜ settlement of suspended sediments
➜ **flocculation**.

If the rate of deposition exceeds the rate of marine erosion the delta will continue to form. As the quantities of deposited sediment continue to grow the channel will become choked and it will be forced to divide into a number of **distributaries** that will grow into one of four characteristic shapes:

➜ cuspate
➜ arcuate
➜ bird's-foot or lobate
➜ estuarine.

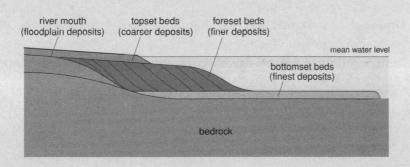

river mouth (floodplain deposits)   topset beds (coarser deposits)   foreset beds (finer deposits)

mean water level

bottomset beds (finest deposits)

bedrock

Exam questions
answers: page 34

1 How might the following changes in the drainage basin affect the growth of a delta: (a) building a dam, (b) deforestation of several slopes? (10 mins)

2 How might the characteristics of a lower river valley be altered by rejuvenation? (10 mins)

3 Describe the characteristic landforms of a river basin that you have studied and explain how they are related to the longer term changes that have taken place in the area. (20 mins)

Checkpoint 2

Use this diagram to help you explain the formation of levées and bluffs.

The jargon

Flocculation is the coagulation of clay particles in suspension when they meet sea water causing them to settle on the bed quickly.

Action point

Draw a series of simple diagrams to illustrate the four different types of delta.

Checkpoint 3

How are the deposits in the topset beds, foreset beds and bottomset beds different from each other, and why?

Links

Managing river basins, page 24.

Examiner's secrets

As with all geography questions, the examiners are always impressed if you can name real places where the things that you are writing about have actually happened.

Managing natural environments

People have been changing the natural environment for thousands of years. More recently it has been realised that some of these changes are damaging the environment and a more sustainable approach is generally used today.

Managing river basins ●●●

Drainage basins are managed for many different purposes:

→ **Water abstraction** – as the population increases there is a growing demand for fresh water for homes and commercial uses. This has been achieved by river abstraction, reservoir construction and groundwater abstraction. More developed societies generally use more water.

→ **Effluent disposal** – more urban developments create an increasing need to use rivers as a means of effluent disposal. Recent EU directives and Environment Agency standards mean that water quality is now carefully controlled in the UK.

→ **Urbanisation** – this has led to large areas of the catchment being covered with hard surfaces that do not allow any infiltration. The water is channelled through drains and flows into streams and rivers very quickly.

→ **Flood control** – as drainage basins have become more developed, particularly on the floodplains of major rivers, there has been an increasing need to build flood control systems. These include:
  → flood storage basins and dams
  → culverts, weirs and sluices
  → reservoirs
  → early warning systems
  → river widening and improved embankments, river diversions
  → upland reforestation.

→ **Drainage of wetlands and marshlands** – to meet increased demands from agriculture and industry.

→ **Deforestation of upland catchments** – due to population pressure.

Many management schemes have been largely engineering schemes, but lessons learnt from previous mistakes and a greater understanding of river processes mean that more environmentally sensitive, holistic and cost-effective schemes are now being adopted.

The consequences of river management

→ River channels are altered.
→ Sediment levels are altered, e.g. clear-water erosion.
→ Discharge patterns are altered – changes in flooding patterns and creation of more low-flow conditions.

The scale of management schemes

River basin management schemes can be multi-purpose, to justify the cost, and may affect the whole of the drainage basin. Examples include the Colorado, the Rhine, the Nile in Egypt and the Yangtze in China. Small-scale schemes also exist in many drainage basins.

Checkpoint 1

Explain how increased levels of urbanisation in a drainage basin will affect the storm hydrograph.

Checkpoint 2

Can you describe an area where upland deforestation as a result of population pressure has resulted in problems in lower parts of the drainage basin?

The jargon

Holistic means 'the whole thing', so here it means consider all aspects of the drainage basin, not just the immediate human needs.

Action point

Can you list examples of each of these consequences from the rivers and drainage basins that you have studied?

Managing coastal environments ●●●

Successful management of coasts involves achieving a balance between conservation and development, resolving conflicts between different users and, vitally, understanding and monitoring the sediment budget. Evaluating the costs and benefits is a complex task. Engineering schemes designed to protect one section of coast may have a detrimental effect further along the coastline.

The sediment budget

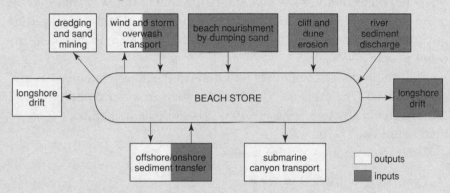

How do coastal changes affect people?

→ Silt deposited in harbours and river mouths requires dredging.
→ Storm waves create damage to coastal settlements.
→ Longshore drift can remove tourist beaches.
→ Increased cliff erosion causes loss of land and properties.
→ Emerging coasts leave ports 'high and dry'.
→ Rising sea levels lead to more coastal flooding.

How do people's actions affect the coastal system?

People use the coastal zone for a variety of purposes and each of these affects the environment in a different way.

Human intervention along a section of coast

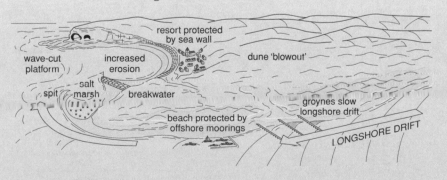

Exam question   answer: page 35

Describe the positive and negative features of a multi-purpose river management scheme that you have studied. (20 mins)

Don't forget

Coastal management is made more difficult because there is no overall coastal planning authority and coastal processes take place across political boundaries.

Action point

Make two lists of the inputs and outputs from the sediment budget diagram to show those that are influenced by human activity and those that are not.

Checkpoint 3

Explain how the management strategies used in this diagram will affect the sediment budget for the area.

Examiner's secrets

It will be vital for you to know detailed case studies to enable you to answer questions on these topics.

Hazard classification and perception

Natural hazards are naturally occurring events or processes that have the potential to cause loss of life or property. They are not simply natural events, as without people they would form no kind of threat. It is the relationship between people and the environment that makes something into a hazard.

Classifying hazards

Why classify? Classifying the many different types of hazards allows people to focus on their key characteristics and in this way they can be more clearly understood. The human response can then be managed by governments, planners, hazard managers, insurance companies and any other people involved.

The simplest classification is into **natural** (such as a hurricane), **quasi-natural** (such as smog), or **human** (such as water pollution). A more commonly used classification is the one described by Bishop in 1998.

The five categories of this classification are:

→ **Type, or geophysical processes**
  - → tectonic → atmospheric
  - → geomorphological → biological
→ **Cause – natural, quasi-natural or human** (see above).
→ **Magnitude and frequency** – the scale of the events and how often they occur. Low-magnitude events are more common than high-magnitude events. Some hazards, e.g. hurricanes, are **seasonal**, whilst lightning and fires are **random**.
→ **Duration of impact and warning time** – there may be a **sudden impact** or a **slow (creeping) onset**, or somewhere in between. It is also important to consider the longer time scale of impacts on human systems as shown in the diagram below.
→ **Spatial distribution** – some hazards only affect certain areas, e.g. hurricanes or tectonic hazards, whereas others are more widespread, e.g. river flooding. There is also a great variation in scale from local to international.

Action point

Can you think of four hazards that fit into each of the three categories of this classification?

Checkpoint 1

What are the scales used to measure the magnitude of earthquakes and hurricanes?

The jargon

Spatial describes distribution over an area. It may be on a large or small scale.

Action point

Use the diagram to describe the main events of a sudden impact hazard that you have studied.

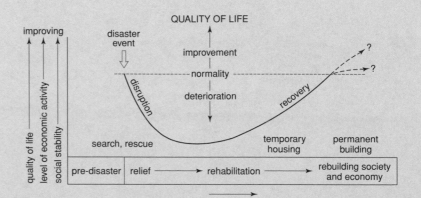

Hazard or disaster?

Some hazardous events take place in areas where the population is not vulnerable so they are not disastrous. In other areas people's lives are adapted to cope with natural processes when they are operating normally, but when events become extreme they may develop into a disaster. The difficulty is in defining a disaster.

→ Monetary values make comparison difficult over long periods.
→ Using numbers of deaths and injuries is a subjective approach.
→ The amount of damage varies greatly depending on the levels of technology and development in the country affected.

Hazard perception

People's perceptions are affected by factors such as their **age**, **sex**, **occupation**, **level of education**, **past experiences** and **expectation for the future**. They may view the threat in various ways:

→ **acceptance** – acts of God, random, beyond our control
→ **domination** – predictable, research, control by technology
→ **adaptation** – natural and human, respond flexibly, research.

Risk

Why do people place themselves at risk from natural hazards? Park (1992) identified a series of factors:

→ unpredictability – time, place and magnitude of event is unknown
→ lack of alternatives – difficult to move due to economic reasons, shortage of land or lack of knowledge
→ changing dangers – once-safe places are now at risk
→ Russian roulette – an optimistic outlook or God's will
→ costs versus benefits – available resources may outweigh the risk.

Vulnerability

Some people are more vulnerable to risks than others. In the Kobe earthquake of 1995, many older and poorer people died as they lived in older and less earthquake-resistant buildings. More wealthy people were also better off after the event because they were able to move away, rather than live in the camps provided for the homeless. In addition to the **wealth** and **technical ability** of people, vulnerability can also be related to **education**, **awareness of the hazard and possible precautions**, **organisation of the society** and **age**. As the world's population is increasing, more people are living in vulnerable areas, particularly in the LEDCs.

Exam questions answers: page 35

1 Explain how perception of a natural hazard may vary according to the level of economic development in the country affected. (10 mins)

2 Explain the difference between a hazard and a disaster, using the example of river floods. (10 mins)

Action point

Carry out research into the definitions of a disaster that are used by various organisations.

The jargon

A *perception* is the way that a person or group of people views a particular situation or event.

Checkpoint 2

Explain each of the three different perceptions of natural hazards by writing out a statement that might have been made by someone with that opinion.

Checkpoint 3

Can you explain why people in LEDCs are not always more vulnerable to natural disasters than people in MEDCs?

Watch out!

Some of the examination syllabuses include human hazards such as disease and crime, but most stick to natural hazards!

Managing hazards

People may perceive hazards in many different ways, as we saw in the last section. These differing perceptions result in different responses to the hazard or different ways of managing the hazard. The range of responses falls into three broad categories which are explained below.

What affects the response?

If a response is made to a hazard it is because there is a willingness and an ability to react having considered any financial, technological or time constraints. There must be **hazard salience**. The type of response made will depend upon a series of factors:

→ the type of event
→ past experiences of hazards
→ economic ability to take action
→ technological resources available
→ knowledge of the available options
→ the social and political framework
→ the perceptions of the decision makers.

Management strategies

These can be corrective (remedial) or preventative. In 1978 Burton, Kates and White identified three groups of options for managing hazards.

1. Modifying the event

The ability to control or change geophysical or biological processes is limited, but some technological solutions such as cloud seeding are becoming more feasible. This strategy can be approached in two ways:

→ **Hazard prevention and environmental control** – flood waters can be controlled by various engineering schemes.
→ **Hazard-resistant design or protection** – building design and construction, e.g. to withstand earthquakes or sea walls.

2. Modifying vulnerability to the hazard

The aims of these responses are to change human attitudes and behaviour towards hazards and in these ways reduce the risks. There are three strategies involved in this type of management response.

→ **Prediction and warning** – this strategy depends upon monitoring of natural events and then the dissemination of the warning followed by an appropriate response from the community.
→ **Community preparedness** – these are measures to prepare the population for the occurrence of the hazard and include education, evacuation procedures and the storage of emergency provisions.
→ **Land-use planning** – the aim here is to prevent any new development taking place in any hazardous areas. To be successful there needs to be considerable knowledge of the potential hazard.

The jargon

Hazard salience is the relative importance of hazards compared with other human issues such as crime, health or poverty.

Action point

Can you list a number of other examples of ways that people have attempted to modify a hazard?

Checkpoint 1

Can you describe three examples of how prediction and warning can modify the vulnerability of a hazard?

Don't forget

Technology and research is always improving, particularly in providing warnings about volcanic eruptions, floods, tornadoes and hurricanes.

3. Redistributing or sharing the losses

There are two ways of sharing the losses following a hazard:

→ **Aid** – provided at community, national or international level:
  → for relief, reconstruction or rehabilitation
  → it may cause political problems, nationally and internationally
  → the UN is now involved in aid work to reduce political problems
  → many charities are involved
  → it is easier to get donations for sudden, high-magnitude events rather than slow-onset events such as droughts.
→ **Insurance** – an important approach in the MEDCs:
  → people pay a series of premiums to cover the cost of any damage
  → the companies spread their losses over a large geographical area
  → in some high-risk areas insurance can be very expensive or may not be available at all
  → as claims increase, insurance will be increasingly difficult to obtain.

Whatever the management strategy adopted the effectiveness will always depend upon the nature of the event, human perception, vulnerability and preparedness. Some strategies will be effective over the long term whilst others will only be effective for a short time. The diagram below is often used to summarise the ways that people make decisions about how to manage hazards.

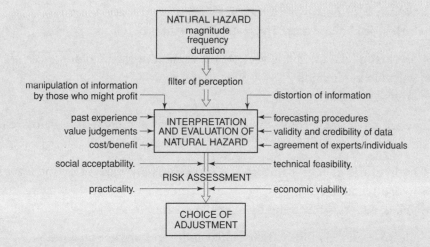

Checkpoint 2

Can you explain an example of a political problem associated with aid as a response to hazard management?

Checkpoint 3

Can you explain why insurance claims are increasing?

Action point

Use the diagram opposite to brainstorm the different points for consideration by a government in response to a predicted volcanic eruption in their country.

Exam questions   answers: page 35

1   For a disaster that you have studied, explain the different ways that the hazard involved has been managed. (20 mins)

2   Using examples that you have studied, describe the types of management that might be possible in attempting to cope with the hazard of frequent river flooding. (15 mins)

Examiner's secrets

To make sure that you can answer any question set it is a good idea to be able to discuss examples of hazards from LEDCs and MEDCs, and also to have knowledge of a variety of different types of hazard and disaster.

Tectonic, geomorphological and atmospheric hazards

Many of the hazards experienced by people can be placed into three categories. **Tectonic** hazards are those related to plate movements and weaknesses. **Geomorphological** hazards are linked with atmospheric and tectonic processes. **Atmospheric** hazards are those related to short- or long-term variations in climatic patterns.

Tectonic hazards

Earthquakes

You need to be able to use the terms **focus**, **thrust fault** and **epicentre** correctly and to be able to explain the main hazards of **ground movement**, and also the secondary hazards of **soil liquefaction**, **landslides**, **tsunamis** (see below) and **avalanches**.

Each earthquake is different in the way that it is hazardous to humans and human activity. You should learn the details of the damage and consequences of your case studies in as much detail as possible. Management strategies include earthquake-resistant engineering, preparation within the community, land use planning, insurance and aid. Currently it is impossible to predict the occurrence of earthquakes accurately, but as technology and research improves this is more likely.

Tsunamis

These are enormous waves that have been created by a large disturbance of sea water. These can be caused by:

→ an earthquake involving vertical movement of the seafloor
→ volcanic explosions
→ large landfalls.

They are most dangerous when they are generated close to a land area. The main effects are:

→ hydrostatic effects – objects and structures are carried by the waves
→ hydrodynamic effects – objects are torn apart and washed away
→ shock effects – battering by material carried in the water.

People are able to modify their vulnerability by predictions and warnings based on earthquake activity, careful planning of land use in vulnerable coastal areas, and building tsunamis-resistant buildings.

Volcanoes

The hazard potential of a volcano depends upon the way that it erupts and the type and amount of material that is released. The various hazards include **lava flows**, **pyroclastic flows**, **ash** and **tephra fall**, **gasses**, **lahars** and **volcanic landslides**. Volcanoes can affect all people in an area irrespective of their social or economic differences. Human responses are usually limited to monitoring, warning and evacuation. However, there have been successful attempts to divert lava by explosions and to cool it down by spraying it with sea water.

Action point

You must be able to describe and explain the location of all these tectonic hazards. You should also be able to explain that they do not only occur at plate boundaries, but also in other locations, and that some are quasi-natural.

Check the net

There are several excellent web sites covering the Kobe earthquake of 1995. Try www.city.kobe.jp/index-e.html

Checkpoint 1

What is the difference between the Mercalli and Richter scales?

Checkpoint 2

Where would you expect to find most tsunamis and why?

Checkpoint 3

Why do you think that tsunamis are more dangerous when they are locally generated?

Check the net

Disaster message service: www.VIEXPO.com/dm/est/earth.html

Example

Study the Mount Pinatubo eruption. It is an excellent example of a large volcanic eruption and shows how prediction and response can be an effective management of a hazard.

Geomorphological hazards

Hazards in the lithosphere result from geological, atmospheric and hydrological processes interacting together.

Landslides

There are many different types of landslide (see page 9).

→ The most dangerous are very fast and involve large amounts of material.
→ Smaller, less dangerous landslides may result in financial losses.
→ Human activity has made the hazard greater by:
　→ building homes and roads on unstable slopes
　→ deforestation of geologically active slopes.

Engineering techniques can stabilise slopes, but who pays and whose responsibility is it to carry out the work? In most cases insurance is not available, and legal liability is the most common way of sharing costs. Forecasting and preparedness can modify vulnerability.

Avalanches

There are three main types: loose snow, slab and slush. They occur mainly on slopes between 25° and 40°. They are hazards only in areas where human activity takes place, and can be managed by controlled avalanches or by deflection devices. Forecasting is improving, many warning systems exist, and land use planning is a mitigating strategy.

Coastal and river flooding

These are frequent natural or quasi-natural hazards because these areas are so attractive for human settlement. River flooding can be caused by high rainfall intensity and totals, snowmelt, glacial outbursts, *jökulhlaup*, landslides and dam failure. The **storm surges** that cause coastal flooding can result from tropical storms, mid-latitude depressions or tsunamis. Flooding can be managed by engineering schemes to control or prevent the floods, warning systems based on prediction and forecasting and preparation among local people.

Atmospheric hazards

Such hazards are usually caused by a short, intense deviation from the normal pattern, but are also related to longer-term climatic changes. El Niño events are closely associated with many of these hazards.

→ **Drought** is a temporary shortage of water supply that is slow to take effect (unlike other hazards).
→ **Tornadoes** – see page 41.
→ **Hurricanes** – see page 40.
→ **Wildfires** are associated with seasonally dry climates or droughts.

Exam question   answer: page 36

Describe the causes, consequences and management strategies associated with a tectonic hazard that you have studied. (20 mins)

The jargon

The *lithosphere* is the earth's crust and rigid upper part of the mantle.

Action point

Find out details about shanty town collapses in Caracas, Cuzco and Rio de Janeiro.

The jargon

A *jökulhlaup* is an Icelandic word for a sudden outburst of meltwater caused by ice melt as a result of volcanic activity.

Checkpoint 4

What is an El Niño event?

Watch out!

Drought is not the same as *desertification* or *aridity*. Aridity is permanent and natural, while desertification is linked to people degrading the ecosystem.

Examiner's secrets

For all these hazards you should be able to describe examples from an MEDC and an LEDC.

Answers
Geomorphology and Hazards

Plate tectonics

Checkpoints

1 Fit of coastlines – Africa and South America; matching geological structures in SE Brazil and S Africa; climatically controlled rock formations – coal in Antarctica; small Permian reptiles found only in Africa and Brazil; palaeomagnetism, sea floor spreading, mid-ocean ridges all along the Mid-Atlantic Ridge.
2 Convection currents in the mantle, which are generated by heat from the centre of the earth, move the solid plates that float on this material. You should add a section about the way that the plates move at each of the three types of plate boundary.
3 Oceanic plate meets continental off west coast of S America where Nazca goes under South American plate. Two oceanic plates converge in Western Pacific. Two continental plates converge in India.

Exam questions

1 North America or Europe are most likely. You should be able to describe young and old fold mountain chains, cratons, geosyncline, and possibly a rift valley.
2 You should describe the way that at least three of the pieces of evidence for plate movement lead people to think that the theory of plate tectonics is the only possible explanation for these phenomena. It is vital to include located examples in your answer.

Earthquakes and volcanoes

Checkpoints

1 At spreading mid-ocean ridges as magma rises between spreading plates. Along continental rift valleys the brittle crust is fractured allowing magma to rise from below. In island arcs and in fold mountain ranges magma rises under pressure created from the melting lithosphere. In isolated locations known as hot spots, volcanoes are due to a rising plume of molten material.
2 Compressional, shear (both high frequency), and Love and Rayleigh (low frequency).

Exam questions

1 Hazardous consequences are to do with the disastrous effects. They include loss of life, damage to property, stress, loss of employment, etc. Advantageous effects include more knowledge of our planet such as better knowledge of mineral sources, rich soils from volcanic ash and lava breakdown, creation of new land, geothermal energy, minerals and tourism.
2 In Japan earthquakes have a deep focus because the subduction zone, where the foci are, is beneath the thick continental plate. Earthquakes associated with the Mid-Atlantic Ridge usually have a shallow focus as the movements are only associated with the shallow ocean plates.

3 The volcanoes of Hawaii are hot spot volcanoes and are the only type not found near a plate boundary.

Weathering and mass wasting

Checkpoints

1 Limestone is chemically weathered by the process of carbonation, i.e. it is dissolved by weak carbonic acid that forms as a result of rainwater absorbing some carbon dioxide and is then removed in solution.
2 Flow is a slow, almost continuous movement downslope of material retaining moisture. A slide is a rapid movement of material along a well-defined plane such as a landslide. Heave is the raising of particles at right-angles to the slope and the subsequent falling down the slope due to the effects of gravity. Rotational slips are a combination of slides and flows.

Exam questions

1 Mechanical weathering requires changes in temperature so it is most efficient in environments where the temperature fluctuates above and below freezing and can operate in dry conditions. Chemical weathering is most effective when water is present and when temperatures are high as, with the exception of carbonation, all chemical processes involved in weathering operate better at higher temperatures.
2 Use examples to show that mass movements have many different characteristics including speed of flow, degree of saturation by water, and type of movement. Each movement is different and has different characteristics of these variables, and this makes it difficult to arrive at one easily applied classification.
3 Choose your rock type carefully and describe the landforms that result. As you describe the feature, explain how it has been formed. For granite you could include tors, core stones, kaolinite and quartz crystals, and for limestone swallow holes, caves, gorges, limestone pavements, dolines and flowstone.

Glacial systems and processes

Checkpoints

1 Ice formation requires alternating periods of freezing and thawing, which do not take place in Antarctica or Greenland.
2 Cold glacier – no melting, gentle gradients, temperatures lower than pressure melting point. Warm glaciers – melting so move faster due to lubricating water, especially in summer, steeper gradients, and base temperatures about same as pressure melting point.
3 Avalanches – sudden fall of ice rock and snow down a steep slope.
Basal sliding – movement of ice over the lubricated valley floor.
Internal plastic flow – plastic deformation of ice as individual particles move past each other under pressure and the influence of gravity.

Exam questions

1 Use the information on the two diagrams on page 10 to help you. You need to describe how the mass of the glacier changes (glacial budget) and relate this to the amount of precipitation and temperatures change between summer and winter, and then relate these to the processes of accumulation and ablation.

2 Global warming is raising the air temperatures so there will be more meltwater, greater evaporation, and more calving of ice sheets, especially where sea level is also rising, leading to a loss of mass. There may be more precipitation as weather patterns change, adding to the mass.

3 Drumlins are smooth elongated mounds of unsorted till, sometimes with a rock core. There is a steep stoss end facing the direction from which the ice came and a gently sloping down-stream lee side. Dimensions may be over 50 metres high and a few hundred kilometres long and wide. They are found singly or in swarms in glaciated valleys.

Glacial landforms

Checkpoints

1 Freeze–thaw action breaks up the rock under the snow and then meltwater removes the loose material as it flows out of the hollow.

2 Terminal – high ridge of unsorted material deposited by the glacier stretching across a valley marking the furthest point reached by the ice.
Lateral – an embankment of frost-shattered debris along the side of a valley carried along by the glacier from its point of origin.
Push – previously deposited moraine material that is shunted into a mound by a temporary ice advance. Some stones may be pushed into an upward position.

3 Eskers – long, narrow, winding ridges of sorted sands and gravel. Formed by deposition from subglacial streams.
Kames – undulating mounds of sand and gravel deposited unevenly along the front of an ice sheet; as the front retreats the mound collapses.
Kame terraces – flat areas of sand and gravel along the valley side. Deposited by streams flowing along the groove between the valley side and the glacier.

Exam questions

1 Roches moutonnées are solid pieces of bedrock that protrude from a glaciated valley floor. On one side they are smooth and gently sloping (stoss side) and on the other are rough, steep and uneven. As a glacier meets a resistant protrusion it slides over the stoss side polishing and scratching it. Then, by regelation (the refreezing of water into ice as the pressure is reduced on the downglacier side of an obstacle), it plucks out rocks that have been loosened by freeze–thaw and pressure release from the ice side.

2 Fluvioglacial deposits are sorted into different sizes, are rounded as a result of being moved by running water and are usually laid down in layers. Directly deposited material is completely unsorted, containing materials of a variety of shapes and sizes and degrees of roundness.

Periglacial processes and landforms

Checkpoint

If mean annual temperature is below –5°C, any melting is extremely rare so continuous permafrost may be 700 metres thick. Discontinuous permafrost is where mean annual temperature is between –1°C and –5°C and is in the form of islands of permanently frozen ground, with less cold areas near lakes and rivers in between. Patches of sporadic permafrost are found where mean annual temperature is just below freezing point and summer temperatures above 0°C.

Exam question

You need to describe how the processes of solifluction work to explain the solifluction lobe, freeze–thaw weathering for the scree, stone sorting, heave and gravity for the stone polygons, and stone stripes and ice accumulation for the pingo.

Coastal processes and landforms

Checkpoints

1 Wave quarrying or hydraulic action is the energy of the breaking wave, abrasion/corrasion is the wearing away of cliffs by other material thrown by waves, attrition is the breaking down and smoothing of loose material.

2 As a headland retreats it might contain geos, caves, arches, stacks and stumps. You need to say how the processes in the previous Checkpoint act to erode the headland.

Exam questions

1 Transitory means short lived. Beaches are changed by every tide to some extent and as there are two tides each day they are truly transitory.

2 Longshore drift has been responsible, in part, for the formation of spits, tombolos and bars. The process is the movement of material along the beach by wave action. Material is moved up the beach at the angle of the swash and is then dragged down the beach by the backwash at right-angles to the beach line.

3 Concordant coastlines include the Dalmatian coastline of Croatia where there are long, thin islands parallel to the coast with narrow channels separating them from the mainland because the rock structure runs along the coastline. A discordant coastline is to be found in south-west Ireland. Here the structure of the rocks in the area runs across the line of ridges and valleys of the coast resulting in many inlets and long headlands. These two types of coastline are more likely to be created following a rise in sea level.

Hydrological cycle and drainage basin systems

Checkpoints

1 When the water balance is negative, water is being lost from the area quicker than it is being replaced by precipitation, so the level of the water table will fall.

2 Regulation of flow. Peaks would be considerably reduced and flow levels would be similar and linked to demand in the basin for water supplies.

Exam questions

1 Here you have to consider that snow, for example, will be lying on the surface for some time before it melts and that the soil may be frozen. Runoff after very intense rainfall is faster than following gentle rain so it will not all penetrate the surface. Continuous rainfall may well fill up the groundwater and soil store after a while and then surface runoff will increase.

2 Due to the leaves on the trees in summer there will be more interception and evaporation, resulting in a much longer lag time. You will need to explain the processes of interception, stemflow and evaporation.

3 This question needs to be broken down into sections to explain each of the differences in turn. Consider small and large drainage basins as opposites. Smaller basins should have quicker response times but relate these to the area that a rainstorm might affect. Consider the difference between lag times in a round basin compared with a long, thin basin. The longer one should have a more 'flashy' response.

River channel and basin processes

Checkpoints

1 Traction moves large, heavy material along the bed. Saltation is the jumping or bouncing of pebbles, sand and gravel along the bed. Suspension is fine particles of clay and silt carried by turbulence in fast-flowing water. Solution is dissolved material.

2 *Competence* is the diameter of the largest particle that a river is able to carry as it moves at a certain speed. *Capacity* is the largest amount of load that a river can carry as it moves at a certain speed.

3 The base level of a river may not be at sea level if the river flows into an inland depression that lies below sea level or into a lake or reservoir normally found considerably above sea level.

Exam questions

1 A positive change in base level means that sea level rises in relation to the land or that the land seems to sink. The result is that the gradient of the river falls and this leads to more deposition and the increased possibility of flooding in coastal areas.

2 The graded profile is a smooth profile attained when the river is in a state of dynamic equilibrium. In the situation

shown the lake will gradually fill up as deposition will be greater than erosion in this part of the system. At the waterfall erosion will be more important than deposition and the feature will eventually retreat and disappear.

Fluvial landforms

Checkpoints

1 Oxbow lakes are formed as a river becomes increasingly sinuous. Erosion on the outside of the bend and deposition on the inside of the bend lead to the neck of the meander becoming increasingly narrower until the main channel cuts through. The section of the former channel becomes silted up and is called the oxbow lake.

2 Levées are the parts of the riverbank that are higher than the rest of the floodplain. They are built up from layers of sediment dropped when the river floods. The coarsest material is deposited closest to the river. Bluffs are steep cliffs along the edge of the floodplain. They are formed as a result of lateral erosion by the river up to the edge of the floodplain and into the surrounding higher land.

3 The bottomset beds are made of the finest material as this is carried furthest. The foreset beds cover the bottomset beds; they are made of coarser material and deposited on a slope. The topset beds are the nearest to the land; they are horizontal and consist of the largest particles.

Exam questions

1 (a) Dam construction will reduce the amount of sediment available to build the delta and if erosion by the sea remains constant the delta may well reduce in size.

(b) Deforestation will reduce the vegetation cover and tend to increase the soil erosion leading to an increase in the load of the river. This will cause an increase in the development of the delta if all other factors remain unchanged.

2 Rejuvenation means a lowering of base level and this leads to greater erosion, particularly near the mouth. Your answer needs to contain reference to knick points, river terraces and incised meanders. You should also mention the way that the extra powers of erosion will be transferred upstream as the processes continue.

Managing natural environments

Checkpoints

1 Increased urbanisation in a drainage basin will result in the building of more hard surfaces and improved drainage systems. This will mean that there will be little infiltration and that precipitation will reach the river channel very much more quickly. The hydrograph will reflect these changes by having a lower base flow and a more 'flashy' shape with a steeper rising limb, shorter lag time and a higher peak flow.

2 The Ganges basin is the best known example of this. In Nepal there has been considerable deforestation as a

growing population has created an increasing demand for wood for fuel and timber and for farmland to feed itself. This has led to increased soil erosion on the slopes and more active river erosion as the sediment load of the river has increased. There has been more flooding in the lower parts of the basin as the rivers have deposited more of this load in the lowland plains.

3   Groynes and breakwaters will trap more sediments. New moorings will protect beach so reducing sediment losses. There will be increased erosion at headlands as more sediment is trapped on the beaches, adding to the sediment budget.

Exam question

Your answer should be divided into two clear sections explaining the positive and negative aspects and **must** focus on a real river that you know and have studied. Positive features might include improved flood control for people living on the floodplain, cheap electricity leading to the development of new industries and therefore better employment opportunities, improved water supply for urban areas and agriculture, etc. The negative aspects might include loss of valuable land as reservoirs are filled, need to borrow capital for the project and thus problems of debt, loss of regular flooding to water the crops, sediment build-up behind dams, clear-water erosion, loss of fishing industry, etc.

Hazard classification and perception

Checkpoints

1   Earthquake vibrations are measured using the Richter scale, and hurricanes by the Saffir-Simpson scale.

2   Acceptance – we can't do anything about these hazards, they are an act of God and just part of everyday life in these areas. Domination – our scientists are able to understand these hazards and by careful engineering solutions we are able to control them. Adaptation – as natural hazards will continue to happen we will have to learn all that we can about them and adapt the way that we live to make sure that we keep their effects to a minimum.

3   Many people in LEDCs live in simple buildings that collapse around them without causing too much loss of life whereas people in MEDCs use concrete and bricks that cause far more injuries when they collapse. People in LEDCs may not be as vulnerable to financial loss as they own little compared with people in MEDCs, however, their livelihood may be completely destroyed.

Exam questions

1   Perceptions of a natural hazard may vary between acceptance, domination and adaptation. In LEDCs there is a lower ability to understand the hazard in detail and a lower ability to undertake technological solutions and therefore they are likely to view a hazard with acceptance or maybe adaptation. Depending upon the individuals or the organisation concerned, people in MEDCs are more able to understand and control hazards and they are likely to view hazards with either domination or control.

2   Living on a fertile floodplain with regular small floods is hazardous because the people know that a flood of larger, damaging proportions is possible. They benefit from the fertility and irrigation water that the river floods bring. The disastrous flood may bring loss of life, crops, property and communication problems.

Managing hazards

Checkpoints

1   Examples of prediction and warning include: (a) the hurricane warnings issued in the USA and cyclone warnings in Bangladesh; (b) flood warning systems on most major rivers in the world; (c) detailed weather forecasts available in most nations; (d) warnings of volcanic eruptions and even earthquakes as a result of seismographic research.

2   Political problems include: (a) civil wars making it difficult for aid to reach the intended recipients in countries like Ethiopia; (b) dissatisfaction when national resources are spent on providing aid for victims who have not taken precautions, such as insurance, against a hazard.

3   Insurance claims are increasing because the population that is living in vulnerable locations is increasing, more people have been able to afford insurance as standards of living have increased.

Exam questions

1   This question requires you to have detailed knowledge of a hazard and to apply the management strategies of: modifying the event, modifying vulnerability and modifying the loss, to an event that you have studied. The best examples might be earthquakes such as Kobe or Hanshin (1995) or a recent one in California. You might also use a volcanic eruption or a drought.

2   These strategies are of the 'modify the event' type: levée or dyke construction, dams and floodwater storage basins, channel management. These strategies are 'modify vulnerability': forecast and warning using modelling, accurate recording equipment and improved communication technology; land-use planning and community preparation schemes are also part of this strategy. 'Modifying the loss' could include insurance. Depending upon the examples that you use, you should refer to as many of these strategies as possible.

Tectonic, geomorphological and atmospheric hazards

Checkpoints

1   The Mercalli scale is a descriptive scale used to describe the intensity of the shaking of the ground during an earthquake. The Richter scale is logarithmic and measures magnitude or total energy release of an earthquake.

2 Tsunamis are usually found on subduction plate boundaries as it is in these locations the sea bed experiences vertical movement that causes most of these events.

3 This is because they move very quickly and there is little time for people to evacuate. Some 99% of deaths resulting from this hazard are caused by locally generated tsunamis. The greater the distance travelled, the longer time there will be to warn and evacuate people.

4 El Niño is a warm ocean current that sometimes replaces the usual cold Peru current off the east coast of South America. It is the reversal of the usual patterns in the area and is linked to changes in the ITCZ in the area. It occurs at intervals of between two and seven years and usually has dramatic effects on climate in several locations around the world.

Exam question

For questions like this it is essential that you know detailed case studies. Here you are asked to describe the causes. You need to refer to tectonic activity/plate movements that were involved in the event. The consequences are the effects the event had on human and natural environment. Include longer-term effects such as loss of employment opportunities or the impact of insurance claims. When discussing the management strategies, refer to the strategies described on pages 28 and 29 and to ensure that you include modifying the event, modifying vulnerability and modifying the loss.

Weather and climate

Weather and climate is vital to the study of physical geography and human geography. Weather and climate are important in their own right but also influence society through agriculture, tourism, hazards, insurance claims, water supplies and so on. Understanding weather and climate is essential for understanding how people manage the environment in which they live. Moreover, in an increasingly urban-industrial world, the experience of weather and climate is most people's only contact with the natural environment.

Exam themes

→ Global circulation patterns

→ Air masses

→ Low pressure systems

→ High pressure systems

→ Weather hazards

→ Contrasting climate (monsoon, tropical & temperate)

→ Small-scale climate e.g. microclimate.

Topic checklist

O AS ● A2

	EDEXCEL		OCR		AQA		WJEC
	A	B	A	B	A	B	
Circulation systems	●	●	O	O●	O	●	●
Weather hazards	●	●	O	O●	O	●	●
Air masses: Lamb's airflow types	●	●	O	O	O	●	●
Characteristics of air masses	●	●	O	O●	O		●
Weather fronts	●	●	O	O	O		●
High pressure systems	●	●	O	O●	O		●
Pollution	●	●	O	O●	O		●
Acid rain	●	●	O	O	O	●	●
Water pollution	●	●	O		O	●	●
The greenhouse effect	●	●	O	O●	O		●
The monsoon	●	●			O		●
The Mediterranean climate	●	●		O	O		●
Temperate climates	●	●		O	O		●
Tropical climates	●	●			O		●
Urban microclimates	●	●		O●	O	O	

Circulation systems

The main factor causing atmospheric circulation is the unequal heating of the earth at different latitudes. In the tropics there is a surplus of energy, whereas polewards there is a deficit of energy. Energy is transferred from low latitudes to high latitudes to balance this unequal heating. Air blows from high pressure to low pressure. The global and local distribution of high pressure tells us which way the winds will blow.

What happens to air in the tropics?

At the equator warm air rises and creates an area of low pressure. The air that has risen at the equator cools down at a higher level. At the subtropics (20–30° North and South) this air sinks because it is colder and denser. (Also, there is less space in the atmosphere with increasing latitude.) This forms high pressure at the subtropics. From here, air returns to the equator to replace the rising air. This convectional cell is known as the **Hadley cell**.

What happens to air in polar areas?

Cold dense air sinks at the poles (the polar high pressure belt), and moves outwards to the mid-latitudes. Between the Hadley cell and the **Polar cell** is another cell, the **Ferrel cell**, which is driven by the movement of the other two cells.

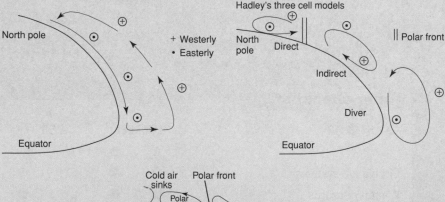

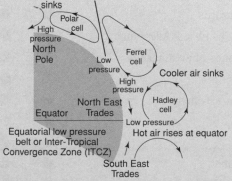

The three-cell model shown above is a useful simplification. However, it *is* very simplified. For example, in addition to the north–south transfers there are:

→ east–west transfers of energy (the low-latitude Walker circulation, and the El Niño reversal)
→ seasonal shifts of the overhead sun, and the position of the thermal equator.

Checkpoint 1

Why does rising air produce low pressure?

Examiner's secrets

Weather and climate is difficult. Students who use diagrams to help explain their answers generally achieve higher grades.

38

New circulation models ●●●

New models change the relative importance of the three convection cells in each hemisphere. These changes are influenced by jet streams and Rossby waves.

Jet streams

These are strong and regular winds that blow in the upper atmosphere about 10 km above the surface. There are two streams in each hemisphere – one between 30° and 50°, the other between 20° and 30° blowing at 100–300 km/h. In the northern hemisphere the polar jet flows eastwards, and the subtropical jet flows westwards.

Jet streams result from differences in equatorial and subtropical air, and between polar and subtropical air. The greater the temperature difference, the stronger the jet stream.

Rossby waves

Rossby waves are 'meandering rivers of air' formed by westerly winds. There are three to six waves in each hemisphere formed by major relief barriers, such as the Rockies and the Andes, and thermal differences between warm air and cold air. Mountains create a wave-like pattern, which typically lasts six weeks. As the pattern becomes more exaggerated (see below) it leads to blocking anticyclones (blocking highs) producing prolonged periods of unusually warm weather in mid-latitudes.

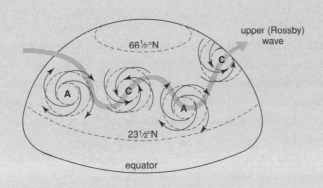

 surface pressure system and winds   **C** mid-latitude cyclones

surface highs/lows   **A** mid-latitude anticyclones

Exam questions   answers: page 68

1 What is the main cause of atmospheric circulation? (5 mins)

2 Why does air rise at the equator but sink at the subtropics? (5 mins)

3 What is a Rossby Wave? (5 mins)

4 What is a jet stream? (5 mins)

Checkpoint 2

What is the *Hadley cell*?

Checkpoint 3

Where are the main global locations of high and low pressure?

Check the net

Look at the Met Office Home page at www.metro.govt.uk

Test yourself

Draw a simple diagram showing jet streams and Rossby waves.

Examiner's secrets

Use diagrams. Too few students use them yet they are excellent tools in an exam. They can help you to explain difficult concepts easily.

Weather hazards

A **hazard** is a natural event that threatens both life and property – a **disaster** occurs when the hazard takes place and human life and property are put at risk. Weather hazards are very varied. They can be natural or man-made, local or global.

Types of weather hazard

→ Fog
→ Snow and ice
→ Droughts
→ Floods
→ Frosts
→ Hail
→ Hurricanes
→ Tornadoes

It is possible to characterise weather hazards and disasters in a number of ways:

1 **Magnitude** – the size of the event.
2 **Frequency** – how often an event of a certain size occurs.
3 **Duration** – the length of time that an environmental hazard exists.
4 **Areal extent** – the size of the area covered by the hazard.
5 **Regularity** – some hazards are regular, such as cyclones, whereas others are much more random, such as tornadoes.

Fog

Fog persists longer when there is a temperature inversion – i.e. when cold air at the surface is overlain by warm air. This is common in high pressure conditions, in valleys and over urban areas. Cold air, being denser, is unable to rise – thus the fog persists. In areas where there are large concentrations of smoke, sulphur dioxide and other pollutants, smog is formed.

Fog is a major environmental hazard – airports may be closed for many days and road transport is hazardous and slow. Freezing fog is particularly problematic. Large economic losses result from fog, but the ability to do anything about it is limited. This is because it would require too much energy (and hence cost) to warm up the air or to dry out the air to prevent condensation.

Tropical cyclones: hurricanes

Hurricanes are intense hazards that bring heavy rainfall, strong winds and high waves, and cause other hazards such as flooding and mudslides. Hurricanes are also characterised by enormous quantities of water. This is due to their origin over moist tropical seas. High-intensity rainfall, as well as large totals (up to 500 mm in 24 hours), invariably cause flooding. Their path is erratic, so it is not usually possible to give more than twelve hours' notice. This is insufficient for proper evacuation measures.

Checkpoint 1

Many people choose to live in hazardous areas. Why?

Checkpoint 2

Why do urban areas experience a high incidence of fog?

The jargon

A *hurricane* is a tropical low pressure system. A *cyclone* is a mid-latitude depression, which is also a low pressure system.

For hurricanes to form a number of conditions are needed:

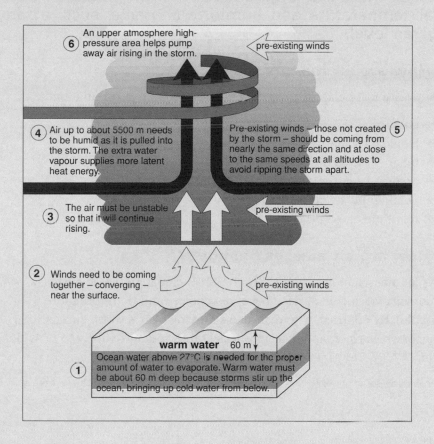

Hurricanes create a major threat to human life, property and economic activities. They are a seasonal hazard, peaking between June and November in the northern hemisphere. Because of their impact, and the cost of their destruction, they are monitored intensely by satellite, and hurricane paths are predicted by complex computer programs.

Tornadoes ●●●

For a tornado to occur in the USA, a number of factors need to occur simultaneously. These include:

→ a northerly flow of marine tropical air (from the Gulf of Mexico) which is humid and has temperatures at the ground in excess of 24°C
→ a cold, dry air mass moving down from Canada or out from the Rocky Mountains at speeds in excess of 80 km/h
→ jet stream winds racing east at speeds in excess of 390 km/h

These three air masses, all moving in different directions, set up shearing conditions, imparting spin to a thundercloud.

Exam questions answers: page 68

1 Why is fog an environmental hazard? (5 mins)

2 (a) How are hurricanes formed?

(b) Why are hurricanes considered to be hazards? (15 mins)

Check the net

Visit the US National Hurricane Center at: nhc-hp3.nhc.noaa.gov

Checkpoint 3

How do tornadoes and hurricanes differ?

Examiner's secrets

You do not need to refer to the latest hazard to score most marks. You need examples that illustrate the many different aspects and impacts of a hazard. Some of these examples may be a few years old.

Air masses: Lamb's airflow types

An air mass is a large body of air with relatively similar temperature and humidity characteristics, at any given level.

How are air masses classified?

Air masses are classified by their surface temperatures and humidity:

→ equatorial (E) are hot
→ tropical air masses (T) are warm
→ polar (P) are cool
→ arctic – from the poles – (A) are cold
→ continental air masses (c) are dry
→ maritime ones (m) are moist.

Checkpoint 1

What is an *air mass*?

How are air masses changed?

Checkpoint 2

How does warm air change when it passes over a cold surface?

As air masses move, their characteristics change. A warm air mass moving over a cold surface becomes cooler (and therefore more stable). By contrast, a cold air mass moving over a warm surface becomes warmer, and therefore less stable and more likely to provide rain.

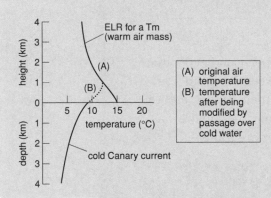

(a) Because the air is chilled, it becomes denser and therefore more stable – often associated with fog.

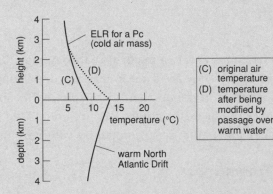

(b) Because the air is warmed, it becomes less dense causing increased instability.

The jargon

The *environmental lapse rate (ELR)* is the decrease of air temperature with altitude.

Air masses originate in source areas – areas of high pressure and high winds. As air blows out from a centre of high pressure, air masses move away from their source regions.

Lamb's airflow types

The geographer Harry Lamb identified seven major categories of airflows (movement of pressure systems) influencing the British Isles. Each airflow type is associated with corresponding air masses:

→ **Westerly** Polar maritime (Pm), Tropical maritime (Tm)
→ **North-westerly** Pm, Arctic maritime (Am)
→ **Northerly** Am
→ **Easterly** Arctic continental (Ac), Polar continental (Pc)
→ **Southerly** Tm or Tropical continental (Tc) – summer Tm or Pc winter
→ **Cyclonic** Pm, Tm
→ **Anticyclonic** Pm, Tm

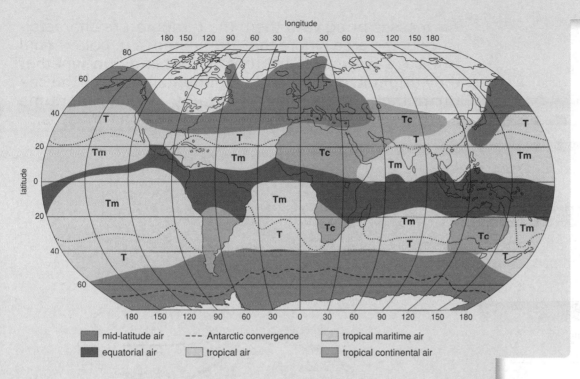

	longitude

Legend:
- mid-latitude air
- equatorial air
- --- Antarctic convergence
- tropical air
- tropical maritime air
- tropical continental air

Type	General weather
Westerly	Unsettled weather with variable wind directions as depressions cross the country. Mild and stormy in winter, generally cool and cloudy in summer.
North-westerly	Cool changeable conditions. Strong winds and showers affect windward coasts especially, but the southern part of Britain may have dry, bright weather.
Northerly	Cold weather at all seasons, often associated with polar lows or troughs. Snow and sleet showers in winter, especially in the north and east.
Easterly	Cold in the winter half-year, sometimes very severe weather in the south and east with snow or sleet. Warm in summer with dry weather in the west. Occasionally thundery.
Southerly	Generally warm and thundery in summer. In winter it may be associated with a depression in the Atlantic, giving mild, damp weather, especially in the south-west, or with a high over central Europe, in which case it is cold and dry.
Cyclonic	Rainy, unsettled conditions often accompanied by gales and thunderstorms. This type may refer either to the rapid passage of depressions across the country or to the persistence of a deep depression.
Anticyclonic	Warm and dry in summer apart from occasional thunderstorms. Cold in winter with night frosts and fog, especially in autumn.

Checkpoint 3

How does cold air change when it passes over a warm surface?

Exam question   answer: page 68

How do air masses influence the weather of the British Isles? (40 mins)

Examiner's secrets

For extra marks go on to explain how air masses may be modified and what this does to the weather.

Characteristics of air masses

Air masses bring with them the temperature characteristics of the area where they originated, e.g. polar – cold tropical – warm. Their humidity depends on whether they originated over land (continental) or over the sea (maritime). Continental air is fairly dry and maritime air is moist. Britain receives a variety of air masses.

Check the net

Visit the BBC Weather Centre at:
www.bbc.co.uk/weather/

Checkpoint 1

What is the difference between *continental* and *maritime* air?

Checkpoint 2

Which air masses bring warm, dry weather?

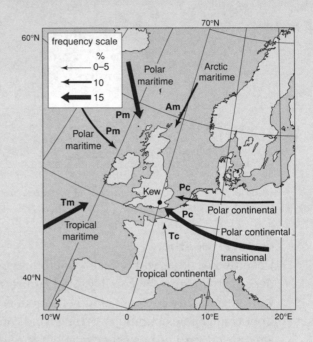

Polar maritime (Pm)

→ An unstable air mass
→ Cool, showery weather – especially in winter
→ Gains moisture over the sea, leading to unstable air
→ 'Nice morning, bad day' cumuliform clouds
→ In winter often about 8°C – as the air mass is warmed by the North Atlantic Drift
→ In summer about 16° – cooled by the North Atlantic Drift
→ The warm North Atlantic Drift encourages convection; visibility is excellent – rising air disperses particles in the air

Arctic maritime (Am)

→ Extreme weather
→ Good visibility

Tropical maritime (Tm)

→ A stable air mass brings warm air from low latitudes
→ Commonly forms warm sector of depressions
→ In winter, air is unseasonably mild (11°C) and damp
→ Stratus or stratocumulus cloud with drizzle – sea fog is common in coastal regions
→ Summer temperatures 16–18°C
→ Visibility poor – solid particles remain near the ground

Tropical continental (Tc)

→ Warmest air entering the British Isles – 13°C in January, 25° in July
→ Can lead to heatwaves or late summer warming – the September 'Indian summer'
→ Can lead to instability and thunderstorms
→ In winter can bring fine, hazy, mild weather
→ Originates in North Africa
→ Moderate visibility – solid particles are not dispersed – and air from the Sahara contains much dust

Polar continental (Pc)

→ Affects British Isles between December and February
→ Very cold, dry air from Siberia around 0°C in January (absent in summer)
→ Picks up moisture from the sea and can lead to snow showers, especially on the east coast (as air becomes unstable over the North Sea)
→ Wind chill factor (dry air) exaggerates coldness

Checkpoint 3

What does Am, Tc, Ac and Pc stand for?

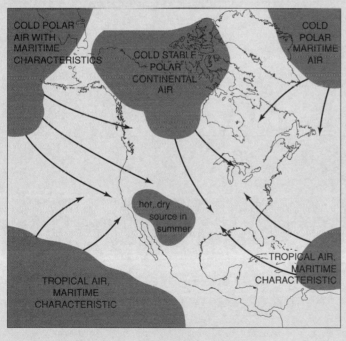

(a) Air masses affecting North America

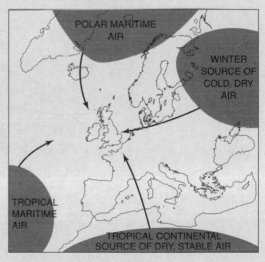

(b) Air masses affecting the British Isles

Checkpoint 4

How do Pm and Tm air masses differ?

Exam question   answers: page 69

With reference to the figures above, discuss why the mid-latitude areas are often described as a 'battleground'. (30 mins)

Weather fronts

The jargon

Low pressure systems are also called cyclones or depressions.

Checkpoint 1

What is a front?

When two different **air masses** meet they form a front. For example, when polar maritime (Pm) and tropical maritime (Tm) air masses converge the temperature differences between them may be over 10°C. This creates differences in density and allows the warmer air mass to rise over the cooler one. In any low pressure system (depression or cyclone) there are a number of forces operating simultaneously:

→ the mixing of the two air masses
→ the Coriolis force and
→ divergence of air aloft in the upper regions of the troposphere.

Low pressure systems

Low pressure systems form over the Atlanic Ocean when moist tropical air meets drier polar air. The warmer, lighter air is forced to rise above the colder air, creating an area of low pressure. The zone where the warm air is rising over the cold air is the warm front.

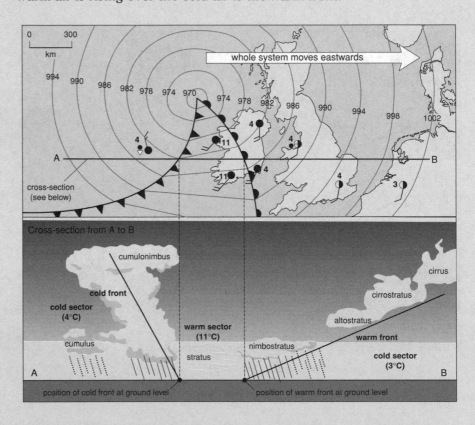

Where the cold air pushes the warm air up, a **cold front** is formed. The rising air is removed at altitude by the jet stream.

In Britain, low pressure systems are more common and stronger in winter.

Checkpoint 2

Why is the pressure in a depression low?

Weather associated with a depression

In general, the appearance of a **warm front** is heralded by high cirrus clouds. Gradually, the cloud thickens and the base of the cloud drops. Altostratus clouds may produce some drizzle, while at the warm front nimbostratus clouds produce rain. A number of changes occur at the warm front:

→ winds reach a peak, are gusty and come from another direction
→ temperatures suddenly rise
→ pressure that had been falling remains more constant.

The cold front is marked by:

→ a decrease in temperature
→ cumulonimbus clouds and heavy rain
→ increased wind speeds and gustiness and another change in wind direction
→ a gradual increase in pressure.

After the cold front has passed, the clouds begin to break up and sunny periods are more frequent, although there may be isolated scattered showers associated with unstable Pm air.

No two low pressure systems are the same. The weather that is found in any depression depends on the air masses involved.

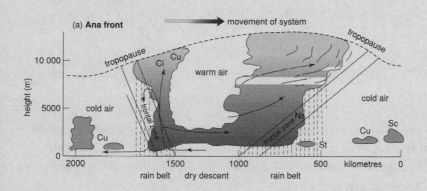

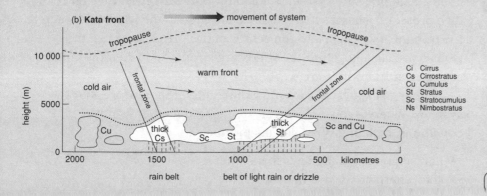

The greater the temperature difference between the air masses involved, the more severe the weather. Depressions are divided into **ana** and **kata** depressions depending upon the vigour of the uplift of warm air. The standard model of a depression was developed by Bjerknes in 1937.

Exam question   answer: page 69

Describe and explain the weather associated with the passage of a low pressure system. (15 mins)

Check the net

An excellent source of satellite images is: www.nottingham.ac.uk/pub/ sat-images/meteosat.html.html

Checkpoint 3

How do warm and cold fronts differ in terms of shape and weather conditions?

The jargon

An *anafront* occurs where air masses of differing composition meet, producing cloud systems of great height. By contrast, *katafronts* occur when the air masses are fairly similar in composition.

Examiner's secrets

Refer to notable storms such as 'The Great Gale' of 1987 – don't worry about being too up to date.

High pressure systems

A high pressure system, or **anticyclone**, occurs when the weather is dominated by stable conditions.

Weather is usually settled and cloudless, but in winter clear skies and light winds can mean frost or fog. Winds blow clockwise around an anticyclone in the northern hemisphere and anticlockwise in the Southern hemisphere.

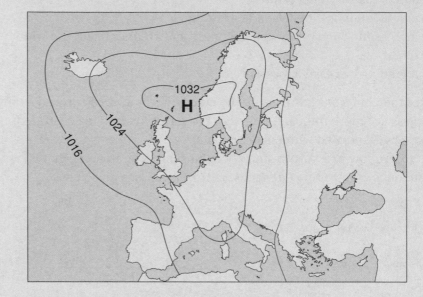

If high pressure persists over northern Europe in winter, then this can mean a spell of very cold east winds for Britain. In summer, however, high pressure over the British Isles or the continent usually brings warm, fine weather.

In a high pressure system winds blow outwards from the centre of high pressure, whereas in a low pressure system the winds blow across the **isobars** at an angle into the centre of low pressure. When the isobars are close together the winds are strong, and when the isobars are far apart the winds are weak.

High pressure systems act very differently from low pressure systems. Whereas low pressure systems produce wet, windy conditions, high pressure systems produce:

➜ hot, sunny, dry, calm days in summer
➜ cold, sharp crisp days in winter
➜ fog and frost in autumn and winter.

The figure opposite shows a weather map for a high pressure system. The weather forecast for London, the South East, Central and North West England for the day stated:

'It will be sunny but cold, becoming wet and windy later. Any fog should clear this morning, leaving plenty of sunshine. There will be a light and variable wind, with temperatures of 2–5°C maximum. Tonight will be frosty with temperatures as low as –5°C.'

The jargon

The H on the weather map means *high pressure* – it does not mean hot!

Checkpoint 1

What is the difference between a cyclone and an anticyclone?

Checkpoint 2

In the northern hemisphere, in what direction does the wind blow in (a) a cyclone and (b) an anticyclone?

Checkpoint 3

Why are nights cold during high pressure conditions in winter?

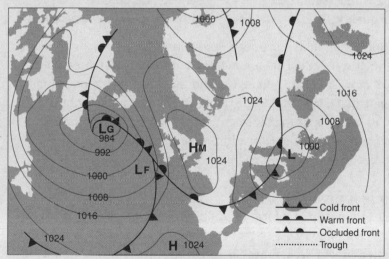

Noon today. Low E will move north-east. Lows F and G will move east and fill. High M will drift east

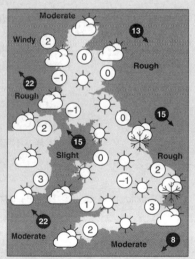

Situation at noon today

answers: page 69

Exam questions

Study the weather maps and the satellite image that shows a high pressure system centred over south-east England.

(a) What is the pressure of this system? (1 min)

(b) Describe the pattern of isobars around the centre of high pressure. What does this tell us about wind speed? (3 mins)

(c) Explain why the day-time and night-time temperatures were so low. (3 mins)

(d) What are the problems associated with high pressure in (i) summer and (ii) winter? (8 mins)

Examiner's secrets

You could refer back to the great high pressure system that dominated the UK in the summer of 1976, which brought hot weather and drought.

Pollution

Pollution is the contamination of the earth/atmosphere such that normal environmental processes are adversely affected. Pollution can be natural, such as from volcanic eruptions, as well as human in origin. It can be deliberate or accidental.

What is pollution?

The levels that constitute 'pollution' can vary. For example, decomposition is much slower in cold environments and so oil slicks pose a greater threat in Arctic areas than in tropical regions. Similarly, levels of air quality that do not threaten healthy adults may affect young children, the elderly or asthmatics.

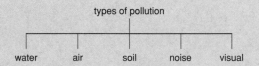

types of pollution

water   air   soil   noise   visual

Pollution leads to:

→ death
→ declining water resources
→ poor air quality

→ decreased levels of health
→ reduced soil quality
→ contamination of ecosystems.

Waste has increased in all European countries. Major sources of waste include agricultural, industrial, municipal and mining activities. The composition of waste is also changing and increasingly includes plastics and packaging materials. In Europe, most waste is disposed of in **landfills**. Without proper management, these can release pollutants into the soil and groundwater. In addition, carbon dioxide, methane and other toxic gases may be produced in landfills.

Managing pollution

It is difficult to develop any form of pollution control when it is easy to estimate the **cost** of controls but more difficult to assess (in monetary terms) the **benefits** of protecting the environment. In addition, it is difficult to assess the actual costs of pollution and to decide who should bear the costs.

It is even more difficult to develop **cross-frontier strategies** when dealing with pollution. The '**ecological time lag**' means that pollution problems are often not recognised until it is too late to do anything about them, let alone decide on a course of remedial action. Unless **point sources** can be targeted it may be impossible to treat pollution effectively. There is no point in treating symptoms, such as acidified lakes with lime, if the cause – the emission of acid materials – is not tackled.

The increase in vehicle exhausts and in sewage waste are notable failures to tackle the pollution problem. By contrast, the reduction of detergent phosphates and the decline of CFCs are good examples of successes.

Checkpoint 1

What are the natural sources of pollution?

Action point

Make a list of the social, economic and environmental costs of pollution, with examples.

Check the net

Visit HMI Pollution at www.open.gov.uk/doe/epsim/inex/html and Manchester air quality at www.doc.mmu.ac.uk.arichome.html

Most synthetic cleaning products come from petrochemicals. Many domestic cleaners contain bleaches and perfumes. Early detergents did not break down rapidly in the environment, but built up in streams and sewage plants (foaming at the surface is characteristic). The production of biodegradable detergents in the mid-1960s reduced the problems, but an additional problem was the use of phosphates to soften hard water and to reduce its acidity. The phosphate-rich waste water accumulated in surface waters and groundwater, leading to eutrophication of streams and lakes. The only way of avoiding this was the development of phosphate-free detergents.

Pollution and economic development

There are a number of views and issues concerning the link between economic and social development and pollution:

→ Is pollution a necessary effect of growth?
→ Is it the price of progress?
→ Are economic development and environmental management two opposing themes?
→ Are they merely a battle between short-term profits and long-term costs?

Although pollution is associated with **capitalist development**, it is not restricted to capitalist countries. The communist countries of the former Eastern bloc have large-scale pollution problems. In particular, the former East Germany has the highest sulphur dioxide emission rates per person in the world. The following methods could reduce the currently unsustainable levels of waste production in most European cities:

→ establish **reduction targets** such as on the emissions of CO_2
→ adopt **waste management plans** to re-use, recycle and recover materials
→ improve **monitoring** of waste sites
→ establish a **comprehensive list** of contaminated sites
→ **coordinate waste management strategies** across international boundaries
→ establish indices of environmental management.

There are certainly more records of pollution in developed countries, but there are increasing levels of pollution in NICs and in developing countries. Countries like Bangladesh have little money to invest in pollution control. Industries in such countries favour the use of cheap, inefficient energy resources, such as lignite and low-grade coal. **Multinationals** are often responsible for pollution in these countries, such as that produced in the Bhopal disaster in India (1984), and the impact of *maquiladora* development in Mexico.

Checkpoint 2
What does the term *sustainable* mean?

Exam question   answer: page 69

'Pollution is the price of progress.' Discuss. (30 mins)

Acid rain

Check the net

Use the Friends of the Earth website www.foe.co.uk/ to find out current information on acid rain.

Checkpoint 1

What is meant by (a) *acid rain* (b) *dry deposition*?

Acid rain – or **acid deposition** – is the increased acidity of rainfall and dry deposition, as a result of human activity. Rain is naturally acidic, owing to carbon dioxide in the atmosphere, with a pH of about 5.6. 'Acid rain' can be a low as 3.0.

Rain has become more than usually acid because of air pollution. Snow and rain in north-east USA have been known to have pH values as low as 2.1. In eastern USA as a whole, the average annual acidity values of precipitation tend to be around pH4. As a general rule, sulphur oxides have the greatest effect, and are responsible for about two-thirds of the problem. Nitric oxides account for most of the rest. However, in some regions, such as Japan and the west coast of the USA, the nitric acid contribution may well be of relatively greater importance. Although emissions of SO_2 are declining, those of NO_x are increasing – partly as a result of increased car ownership.

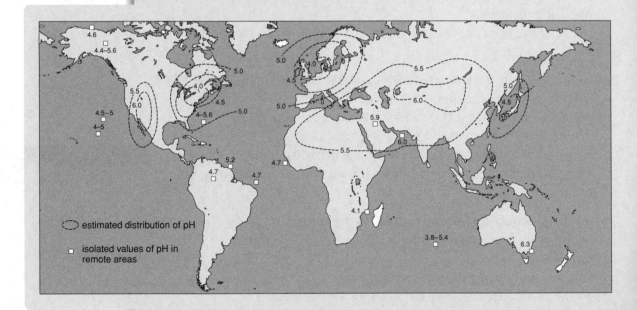

estimated distribution of pH

isolated values of pH in remote areas

Checkpoint 2

What are the following chemicals: SO_2 and NO_x?

Checkpoint 3

What is meant by a logarithmic scale?

The **pH scale** is used as a measure of acidity or alkalinity: 7 is neutral, less than 7 is acidic and more than 7 is alkaline. The pH scale is **logarithmic**, so a decrease of one pH unit represents a tenfold increase in acidity. Thus pH4 is ten times more acidic than pH5.

The major causes of acid rain are the sulphur dioxide and nitrogen oxides produced when fossil fuels such as coal, oil and gas are burned. Sulphur dioxide and nitrogen oxides are released into the atmosphere where they can be absorbed by the moisture and become weak sulphuric and nitric acids, sometimes with a pH of around 3. Most natural gas contains little or no sulphur and causes less pollution.

Coal-fired power stations are the major producers of **sulphur dioxide**, although all processes that burn coal and oil contribute. Vehicles, especially cars, are responsible for most of the **nitrogen oxides** in the atmosphere.

Dry deposition typically occurs close to the source of emission and causes damage to buildings and structures. **Wet deposition**, by contrast, occurs when the acids are dissolved in precipitation, and may fall at great distances from the source. Wet deposition has been called a 'trans-frontier' pollution, as it crosses international boundaries.

Acidification has a number of effects:

→ buildings are weathered
→ metals, especially iron and aluminium, are mobilised by acidic water, and flushed into streams and lakes
→ aluminium damages fish gills
→ forest growth is severely affected
→ soil acidity increases
→ there are links (as yet unproven) with the rise of senile dementia.

The effects of acid deposition are greatest in those areas that have high levels of precipitation (causing more acidity to be transferred to the ground) and those that have base-poor (acidic) rocks that cannot neutralise the deposited acidity.

The solutions

Various methods are used to try to reduce the damaging effects of acid deposition. One of these is to add powdered limestone to lakes to increase their pH values. However, the only really effective and practical long-term treatment is to curb the emissions of the offending gases. This can be achieved in a variety of ways:

→ by reducing the amount of fossil fuel combustion
→ by using less sulphur-rich fossil fuels
→ by using alternative energy sources that do not produce nitrate or sulphate gases (e.g. hydro- or nuclear power)
→ by removing the pollutants before they reach the atmosphere.

However, while victims and environmentalists stress the risks of acidification, industrialists stress the uncertainties. For example:

→ rainfall is naturally acidic
→ no single industry/country is the sole emitter of SO_2/NO_x
→ different types of coal have variable sulphur content
→ cars are also a source of this type of pollution.

Checkpoint 4

Why do some industrialists deny that acidification is due to industry?

Exam questions   answers: page 70

1   Describe the pattern of the most acidified areas in the figure on page 52. (10 mins)

2   Explain the causes of acid rain. (20 mins)

Examiner's secrets

Support your answers with facts, figures and case studies rather than general points.

Water pollution

Water pollution is a 'serious **ecological disaster** comparable in importance to the destruction of the tropical rainforests and desertification'.

Increases in freshwater pollution

There is widespread pollution by sewage, nutrients, toxic metals, industry and agricultural chemicals; the most widespread is from domestic sewage. Poor waste-water treatment and inadequate sanitation have resulted in an exponential increase in waste pollution. Water pollution in many countries intensified in the twentieth century due to:

→ industrialisation
→ urbanisation
→ deforestation for urban growth and agriculture
→ the damming of rivers
→ destruction of wetlands
→ mining and industrial development
→ agricultural development
→ increased energy consumption.

There are six major problems facing the world's fresh water:

→ eutrophication
→ acidification
→ toxic contamination
→ decline of water levels
→ accelerated siltation
→ extermination of ecosystems and biota.

There is some evidence that developed countries have passed through a number of **stages of water pollution**. Developing countries (LEDCs) have experienced fewer stages but are expected to follow suit.

Checkpoint 1

What are the main sources of water pollution?

Checkpoint 2

What does *siltation* mean?

The jargon

Eutrophication means nutrient enrichment.

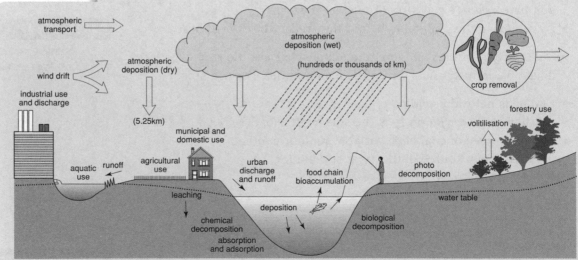

54

The main pollutants ●●●

Pathogens

The most common source of pathogens is **organic matter** from:

→ domestic sewage and municipal waste
→ industrial effluents, e.g. tanneries, paper mills and textile factories
→ storm runoff
→ land fills
→ agricultural areas.

This organic matter includes faecal material, viruses, bacteria and other organisms, as well as a wide variety of carbon compounds. Water-borne infections include schistosomiasis, hepatitis A and dysentery.

In the USA, 16 billion disposable nappies are dumped in landfill sites each year, and are a major source of concentrated pathogens.

Nutrients and eutrophication

Large concentrations of inorganic nutrients such as fertiliser can overload natural systems. In the UK, a major problem results when nitrates from agricultural areas percolate into the groundwater. Many European rivers have extremely high levels of nitrogen and phosphorus – up to fifty times the natural background levels. This overloading causes **eutrophication**. This can create algal blooms leading to oxygen depletion and a decline in biological diversity. World fertiliser use is increasing rapidly, especially in developing countries. High levels of nutrient enrichment are also caused by seepage of water from septic tanks and pit latrines.

Heavy metals

Pollution by heavy metals occurs in a number of ways:

→ processing of ores and metals
→ industrial use of metal compounds
→ leaching from domestic and industrial waste dumps
→ mine tailings
→ contaminated bottom sediments
→ lead pipes.

The effect of heavy metals tends to be regional or local rather than global. However, they can affect areas downstream or downwind; for example, the Rhine Basin supports 40 million people and 20% of the world's chemical industry. Until the 1970s the Rhine was severely polluted but it has improved due to better waste-water treatment and the replacement of certain metals in industrial processes.

Check the net

Visit the World Health Organisation at:
www.who.ch/

Examiner's secrets

A contrast between a developing country and a developed country, with some reference to the model of development and pollution, will improve your chances of success.

Exam question answer: page 70

What are the causes and consequences of freshwater pollution? (30 mins)

The greenhouse effect

The jargon

The lower atmosphere acts like a *greenhouse*, allowing short-wave radiation in, but trapping outgoing long-wave radiation.

The **'greenhouse effect'** is the process by which certain gases absorb outgoing long-wave radiation from the earth and return some of it back to earth. In all, greenhouse gases raise the earth's temperatures by about 33°C.

Properties of greenhouse gases

Greenhouse gases vary in their abundance and contribution to global warming.

Greenhouse gas	Average atmospheric concentration (ppm)	Rate of change (% per annum)	Direct global warming potential (GWP)	Lifetime (years)	Type of indirect effect (ppm)
Carbon dioxide	355	0.5	1	120	None
Methane	1.72	0.6–0.75	11	10.5	Positive
Nitrous oxide	0.31	0.2–0.3	270	132	Uncertain
CFC 11	0.000 255	4	3400	55	Negative
CFC 12	0.000 453	4	7100	116 months	Negative
CO					Positive
NO$_x$					Uncertain

Check the net

Visit a site on the greenhouse effect at: www.dar.csiro.au/pub/info/greenhouse.html

The rise of greenhouse gases

Carbon dioxide levels rose from about 315 ppm in 1950 to 355 ppm in 1999 and are expected to reach 600 ppm by 2050. The increase is due to human activities:

→ burning **fossil fuels** – coal, oil and natural gas
→ **deforestation** of the tropical rainforest is a double blow – not only does it increase atmospheric CO_2 levels, but it also removes the trees that convert CO_2 into oxygen
→ **methane** is the second largest contributor to global warming – cattle convert up to 10% of the food they eat into methane, and emit 100 million tonnes of methane into the atmosphere each year
→ **chlorofluorocarbons** (CFCs) are synthetic chemicals that destroy ozone and absorb long-wave radiation – CFCs are increasing at a rate of 6% per annum, and are up to 10 000 times more efficient at trapping heat than CO_2

Evidence for the greenhouse effect is the link between CO_2 levels and temperatures (modern records only began in 1860):

→ average temperatures have risen by over 0.5°C over the last century
→ July 1998 was the hottest month ever recorded
→ the last decade is the warmest on record
→ nine of the ten warmest years in West Europe have occurred since 1983.

Checkpoint 1

Why is the build-up of methane and CFCs of greater concern than the build-up of carbon dioxide?

The effects of global warming

→ Sea levels will rise, causing flooding in low-lying areas such as the Netherlands, Egypt and Bangladesh.
→ There will be an increase in storm activity (because there will be more atmospheric energy).
→ Agricultural practices and patterns will change.
→ There will be a different distribution of rainfall.

Ways of reducing greenhouse gas emissions

The energy sector

→ Introduce a carbon tax on electricity generation
→ Introduce a higher CO_2 tax and retain the existing energy tax on non-energy-intensive industry

The transport sector

→ New rules and taxes on company cars (to reduce long-distance travel)
→ Expand public transport systems
→ Set CO_2 emission limits on light vehicles

Other greenhouse gases

→ Reduce agricultural use of nitrogen fertilisers
→ Expand methane extraction from waste tips

The ozone hole

Ozone (O_3) is continuously created and destroyed in the atmosphere. Oxygen (O_2) is broken down into individual atoms by ultraviolet radiation. Some of these atoms combine with oxygen to form ozone. Ozone is broken down by ultraviolet radiation – so there is a natural cycle of growth and decay.

Ozone is important as it filters out harmful **ultraviolet (UV) radiation**. However, when CFCs are broken down in the atmosphere they release chlorine, which destroys ozone. As this destruction of ozone is faster than its natural regeneration, the amount of ozone is decreasing.

The effects of reduced O_3 include:

→ increased risk of skin cancer
→ more eye diseases such as cataracts
→ crop yields will decline by 25% (if O_3 declines by 25%).

The ozone 'hole' is a large area over Antarctica (and to a lesser extent over the Arctic) where there are less than half normal levels of O_3.

Checkpoint 2

How is ozone broken down and re-formed?

Exam questions   answers: page 70

1   What is the 'greenhouse effect'? (10 mins)

2   How could global warming increase storm frequency and intensity? (10 mins)

The monsoon

The **monsoon** is the reversal of pressure and winds that gives rise to a marked seasonality of rainfall over north and south-east Asia. This can be seen clearly in India and Bangladesh.

The monsoon has sometimes been described as a giant **land–sea breeze** – but this is too much of a simplification. A number of influences have been suggested:

➜ differential heating and cooling of land compared with the adjacent seas
➜ seasonal movement of the **inter-tropical convergence zone** (ITCZ)
➜ the effect of the Himalayas on the ITCZ
➜ reduced CO_2 due to the Tibetan Plateau (no trees) leads to cooling in the interior.

Winter monsoon ●●●

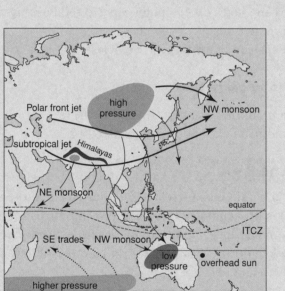

➜ Temperatures over central Asia are low, leading to **high pressure**.
➜ The **jet stream** splits into two: the southern subtropical jet leads to descending air and high pressure.
➜ This brings outward-blowing north-easterly winds across south Asia.
➜ These dry air streams produce clear skies and sunny weather over most of India (November–May).
➜ Bombay receives less than 100 mm during this period.
➜ An intense **low pressure** area develops over northern Australia, where it is summer and very warm.
➜ Winds blow from the Asian high pressure area to the more intense Australian low pressure area.

Check the net

One of the best sites is:
www.geog.le.ac/cti/

Checkpoint 1

What does the term *monsoon* mean?

Checkpoint 2

What is a *land–sea breeze*?

Checkpoint 3

What is a *jet stream*?

Summer monsoon

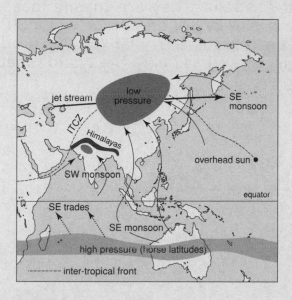

→ In March and May the winds shift, and the upper westerly air currents begin to move north.

→ The **jet stream** strengthens until it lies entirely to the north of the Himalayas.

→ The overhead sun migrates north to a position just over India, and the **ITCZ** moves north (the monsoon trough).

→ Intense low pressure develops over Asia. This is separated, by the Himalayas, from a smaller area of intense low pressure over the Punjab.

→ Strong convectional heating under clear skies leads to low pressure drawing in warm, moist air (Equatorial maritime Em) from over the Indian Ocean.

→ Winds bend from SE to SW on crossing the equator; they then revert to south-easterlies drawn by the Asian low pressure.

→ High pressure develops over northern Australia, where it is winter; winds blow from the Australian high pressure area to the more intense Asian low pressure.

→ Mumbai receives over 2000 mm of rain and Cherrapunji over 11 000 mm in just 4 months. Climate data for Cherrapunji is given below.

	J	F	M	A	M	J	J	A	S	O	N	D
Temp. (°C)	11	13	15	17	18	19	20	21	20	19	16	11
Rain (mm)	24	38	170	595	1700	2900	2700	1700	1200	420	40	10

Action point

Plot the climate data shown in the table and describe the climate experienced at Cherrapunji.

Examiner's secrets

Quote figures. Use the data in the table to show how much rain falls in the rainy season, and how much in the dry season. The top candidates will convert this to a percentage of the year's total.

Exam question   answer: page 70

Describe and explain the weather patterns associated with the Indian monsoon. (30 mins)

The Mediterranean climate

Mediterranean climates occur in many places: in and around the Mediterranean, California, Chile, South Africa and Australia. They are mostly located on the western margins of continents between 30° and 40°. Mediterranean areas are vital for the production of citrus fruits, as well as being the world's leading wine producing areas.

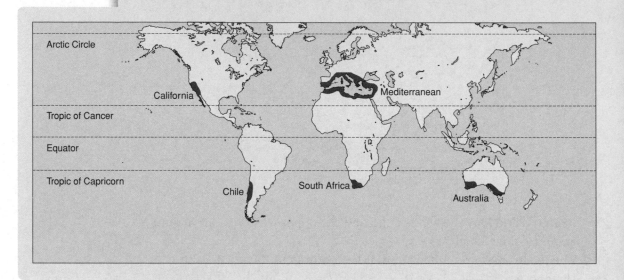

Main patterns

→ The typical climate is on a warm western continental margin.
→ Summer temperatures are over 26°C, and winter temperatures are over 6°C.
→ Rainfall occurs mostly in winter; there are long, dry, hot sunny summers and drought is common.
→ Temperatures are never too low for plant growth, but growth is checked by drought in summer, and winter temperatures are too low for vigorous growth. Hence maximum growth is in the spring and autumn. There is an abundance of spring flowering plants.

Winds

Mediterranean climates are influenced by westerly winds and trade winds. In the summer they are affected by eastern (**continental**) influences and are therefore dry, whereas in winter they are influenced by westerly (**maritime**) air and are therefore wet.

East coast and west coast Mediterranean climates

There is a major difference between west coast Mediterranean and east coast Mediterranean climates. East coast climates are wet in summer and dry in winter, whereas west coast climates are dry in summer and wet in winter.

Checkpoint 1

Why do Mediterranean climates receive most of their rainfall in the winter?

The essential characteristics are a **drought** and a **wet period**. The wet and dry seasons vary with latitude. The rainy season lengthens polewards. Rainfall also decreases eastwards. For example, Gibraltar receives 910 mm, Athens 410 mm.

Average annual temperature ranges and average daily temperature ranges are high. Many Mediterranean areas are noted for their **local winds**, such as the *berg* wind in South Africa, the *bora* in Croatia, and the *mistral* in southern France.

Checkpoint 2

How does rainfall vary in Mediterranean areas? Use examples to support your answer.

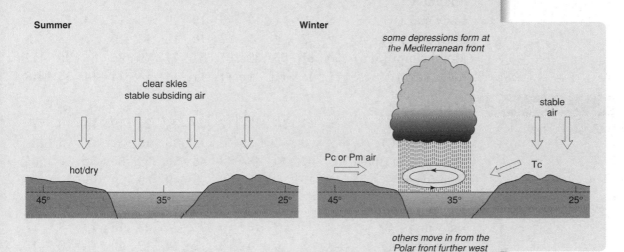

Seasonal contrasts in high and low pressure ●●●

In summer, subtropical high pressure (STHP) dominates the area, creating weak winds and calm conditions. Many areas also have cold offshore currents. This leads to cold air masses that produce limited rainfall. For example, 90% of the winds in California and Chile in the summer come from the sea.

In winter, however, Mediterranean areas experience frontal activity and with it associated rainfall (see the data below for Haifa, Israel).

	J	F	M	A	M	J	J	A	S	O	N	D
Temp. (°C)	14	14	16	19	23	25	28	28	27	24	21	16
Rain (mm)	180	145	23	18	3	0	0	0	0	13	69	170

Exam question   answer: page 71

Sketch the climate data for Haifa in Israel, then describe the main patterns of the climate there. (15 mins)

Examiner's secrets

Draw climate graphs in essays. It saves time in describing the climate and acts as a prompt to help you recall data.

Temperate climates

Temperate climates are those that are mild: not too hot like tropical or equatorial ones, and not too cold like arctic or sub-arctic ones.

Check the net

An excellent site for investigating Britain's weather is: www.meto.govt.uk/

Checkpoint 1

Why does Shannon have a wetter climate than London?

Temperate climates in Europe ●●●

Selected data for European sites are given below.

	J	F	M	A	M	J	J	A	S	O	N	D
Shannon												
Daily max.(°C)	8	9	11	13	16	19	19	20	17	14	11	9
Daily min.(°C)	2	2	4	5	7	10	12	12	10	7	5	3
Monthly ppt (mm)	94	67	56	53	61	57	77	79	86	86	96	117
Rain (days)	15	11	11	11	11	11	14	14	14	14	15	18
London												
Daily max.(°C)	6	7	10	13	17	20	22	21	19	14	10	7
Daily min.(°C)	2	2	3	5	8	11	13	13	11	8	5	3
Monthly ppt (mm)	52	47	40	48	48	44	56	54	50	52	56	53
Rain (days)	11	9	8	8	8	8	9	9	9	9	10	9
Berlin												
Daily max.(°C)	2	3	8	13	19	22	24	23	19	13	7	3
Daily min.(°C)	–3	–3	0	4	8	12	14	13	10	6	2	–1
Monthly ppt (mm)	43	40	31	41	46	62	70	68	46	47	46	41
Rain (days)	11	9	8	9	9	9	11	9	8	9	10	9

Europe's complex and fragmented geography provides a variety of climatological and meteorological conditions. It contains a range of climatic regimes. In general there is:

→ north (colder) south (warmer) variation
→ east–west variation, with places further east being drier and experiencing greater annual temperature ranges than maritime locations in the west.

This gives rise to a four-fold division of European climates including:

→ Mediterranean conditions in the south
→ temperate oceanic in the west
→ temperate continental in the east
→ boreal in the north.

This simple pattern is further complicated by the presence of mountain ranges, proximity to the sea – in particular the North Atlantic Drift – and aspect.

The effect of oceans ●●●

The influence of the ocean is most marked in parts of northern Europe. Areas beyond 45° North (Ireland, Britain and Scandinavia) experience positive temperature anomalies in winter. During summer, the ocean cools these places and they experience a negative temperature anomaly.

By contrast, continental areas such as Moscow are much hotter during summer but colder during winter.

The effect of ocean currents is most noticeable in winter. By contrast, during the summer the ocean has a cooling effect and there is increasing continentality towards the east. Temperature ranges vary from 8° to 10°C in Iceland to up to 28°C in parts of Russia.

Rainfall ●●●

Rainfall levels are highest in western areas and over mountainous regions. This is because most of the rain-bearing winds come from the west, combined with the effect of relief on condensation and precipitation. Rainfall varies from between 1000 and 2000 mm on the west coast of the British Isles to below 500 mm in parts of Sweden, southern Spain, Greece and the Baltic States.

Southern Europe has a dry season (spring and summer) that lengthens eastwards and southwards. By contrast, the rest of Europe receives precipitation throughout the year, with western areas receiving a maximum in autumn and winter. Related to continentality, the proportion of summer rainfall (compared with mean annual rainfall) increases eastwards. In mountainous areas and the interior lowlands of Europe's precipitation is most plentiful during spring.

Even throughout Britain there are significant regional differences. Oceanic climates such as Devon and Cornwall have annual fluctuations in temperature of less than 16°C, mild winters, autumn and winter maxima of precipitation and moderately warm summers; while sub-oceanic climates have an annual temperature fluctuation of between 16° and 25°C, mild to moderately cold winters, autumn to summer maximum of precipitation and moderately warm summers (for example, Gatwick would fit this description).

Checkpoint 2

What is meant by *continentality*?

Exam question   answer: page 71

Refer to the data for Shannon, London and Berlin.

(a) Describe and explain the difference in temperatures in July and January.
(b) Why does Berlin have heavier precipitation in the summer months than London?
(c) Why does Shannon very rarely have temperatures below freezing?
     (10 mins)

Examiner's secrets

Use graphs to show variations in climate. This helps you to describe similarities and differences more easily.

Tropical climates

One of the most enduring and consistent aspects of the global circulation is the tropical climatic system. Rainforest and savanna are both types of tropical climates. Savanna areas are very diverse in the characteristics of their climate – but all have a dry season.

How tropical climates are formed

The overhead sun heats the ground, leading to **convectional rise** of air. This leads to the creation of the **Hadley cells**. These are direct convectional cells where the warm air rises, creating low pressure at the surface.

The air diverges at the tropopause and moves towards the poles until the colder air sinks to the surface. This creates high pressure at 25–40° (the **subtropical high pressure systems** (STHP)). Air from these sub-tropical high pressure belts returns at the surface to the equator as the trade winds. Where these trade winds meet is known as the **inter-tropical convergence zone** (ITCZ). Here the air rises again – this gives the consistency and near-predictability of equatorial weather.

Modification of air

Where the air has travelled over warm tropical oceans it has been modified in its lower layers and becomes **unstable**. It will also have picked up a lot of moisture. Most of the rainfall in the tropics is convectional and the convection that occurs at the ITCZ gives the initial uplift for this unstable air. The characteristic clouds are cumulo-nimbus with associated heavy thunderstorms. The ITCZ moves with the overhead sun.

Rainforest climates

The rainforest has one of the most predictable and stable climates on earth. Rainforest areas have limited seasonal contrasts, as shown by comparison with the data for the Nigerian savanna opposite.

Savanna climates

Savanna areas are very diverse in the characteristics of their climate – but all have a **dry season**, if not two dry seasons; these can vary from as little as one month to as much as eight months. The dry season begins when the sun is near its lowest noon elevation over the horizon.

The **wet season** occurs when the sun is at its highest point and beginning to retreat. The pattern of rainfall follows the ITCZ, which moves with the overhead sun, though over a smaller latitudinal range. The ITCZ represents the area where the two rising arms of the Hadley cells of each hemisphere meet at the surface before rising.

Rainfall in the tropics is predominantly convectional, so it follows that the period of maximum rainfall should coincide with the time when the sun is at its highest elevation, and that the broad rainbelts should follow the overhead sun. The wet season occurs in summer: heavy convectional (monsoonal) rain replenishes the parched vegetation and soil.

Checkpoint 1

What factor characterises all savanna areas?

Checkpoint 2

How do rainforest and savanna climates differ?

The annual range of temperatures is over 10°C in the interiors but less than 4°C in the maritime areas. Annual precipitation ranges from about 250 mm to 2000 mm, enough to support deciduous forest. Two temperature peaks may occur. Savanna areas come under the influence of the trade winds and the doldrums.

Data that show climatic conditions in the tropical rainforest (humid tropics) and the savanna (tropical wet/dry) are given below.

	Tropical rainforest (Amazon, Brazil)		Savanna (Kano, Nigeria)	
Month	Ppt (mm)	Temp. (°C)	Ppt (mm)	Temp. (°C)
Jan.	262	26	0	23
Feb.	196	27	2	24
Mar.	254	26	13	27
Apr.	269	26	64	28
May	305	26	150	27
June	234	26	180	25
July	223	25	216	24
Aug.	183	26	302	23
Sep.	132	27	269	24
Oct.	175	27	74	25
Nov.	183	27	2	24
Dec.	264	26	0	23
Total	2680 mm		1272 mm	

Checkpoint 3

What is convectional rain? What other types of rain are there?

Check the net

Look up www.nottingham.ac.uk/pub/sat-images.meteosat.htmlhtml for up-to-date satellite images and information http://covis.atmos.uiuc.edu/guide/guide.html is an online guide to meteorology, covering many aspects of weather such as fronts, cyclones, pressure, clouds.

Exam questions answers: page 71

1 Describe the climate associated with rainforest areas as shown by the data given above. (15 mins)

2 Describe and explain the climate associated with savanna regions as shown by the data for Kano, Nigeria. (20 mins)

Examiner's secrets

Use information from the net to increase the amount of detail in your answers. The sites suggested provide good access to the information you need.

Urban microclimates

Don't forget!

Urban climates are only noticeable during periods of high pressure. During times of low pressure, winds mix the air above urban areas with that of surrounding rural areas.

The jargon

Albedo means the reflectivity of a surface.

Checkpoint 1

Why is temperature higher in urban areas than in rural areas?

Checkpoint 2

Why is thunder more common in urban areas?

Check the net

Look up www.wcceh.gov.uk/ for details on urban climates.

Urban climates in temperate latitudes differ quite distinctively from nearby rural ones. This is due to a number of reasons.

Structure of the air above the urban area

→ More dust in urban areas means an increased concentration of **hygroscopic particles**
→ Less water vapour
→ More **CO_2**
→ Higher proportions of noxious fumes from the combustion of fuels
→ Waste gases discharged by industry

Structure of the urban surface

→ More heat-retaining materials with lower **albedo** and better radiation-absorbing properties
→ Rougher surfaces with a great variety of perpendicular slopes facing different aspects
→ Tall buildings can be very exposed
→ Deep streets are sheltered and shaded

Resultant processes

1 Radiation and sunshine

There is greater scattering of **shorter-wave radiation** by dust, but much higher absorption of longer waves owing to building surfaces and CO_2. Hence there is more diffuse radiation, with considerable local contrasts, because of variable screening by tall buildings in narrow, shaded streets. Reduced visibility arises from **industrial haze**.

2 Clouds and fog

There is a higher incidence of thicker cloud cover in summer and radiation fogs or **smogs** in winter because of increased convection and air pollution. Hygroscopic particles accelerate onset of condensation.

3 Temperatures

Greater heat energy retention and release, including fuel combustion, gives significant temperature increases from suburbs into the centre of built-up areas, creating **heat islands** (see opposite). These can be up to 4°C warmer during winter nights. Snow in rural areas increases its **albedo**, thereby increasing the differences. Heating from below increases air mass instability overhead, notably during summer afternoons and evenings. There can be strong local contrasts between sunny and shaded surfaces, especially in the spring.

4 Pressure and winds

Severe **gusting and turbulence** around tall buildings causes local pressure gradients between windward to leeward walls. Deep, narrow streets can be calmer unless aligned with prevailing winds to funnel flows along them – the **canyon effect**.

5 Humidity

Decreases in relative humidity occur in inner cities owing to lack of available moisture and higher temperatures. Partly countered in very cold, stable conditions by early onset of condensation in industrial zones.

6 Precipitation

In urban areas there can be more intense storms, because of greater **instability** and stronger convection above built-up areas. Snowfalls are lighter and briefer in urban areas.

The urban heat island

Temperatures 2–4°C higher in urban areas, creating urban heat island. Data shows average changes in climate caused by urbanisation:

Factor	Comparison with rural environments	
Radiation	In general	2–10% less
	Ultraviolet, winter	30% less
	Ultraviolet, summer	5% less
	Sunshine duration	5–15% less
Temperature	Annual mean	1°C more
	Sunshine days	2–6°C more
	Greatest difference at night	11°C more
	Winter maximum	1.5°C more
Frost-free season		2–3 weeks more
Wind speed	Annual mean	10–20% less
	Gusts	10–20% less
	Calms	5–20% more
Relative humidity	Winter	2% less
	Summer	8–10% less
Precipitation	Total	5–30% more
	Number of rain days	10% more
Snow days		14% less
Cloudiness	Cover	5–10% more
	Fog, winter	100% more
	Fog, summer	30% more
	Condensation nuclei	10 times more
Gases		5–25 times more

Exam question   answer: page 71

Describe and explain the main characteristics of an urban climate. (30 mins)

Checkpoint 3

Explain why urban areas are warmer than nearby rural areas, especially at dawn during high pressure conditions?

Checkpoint 4

Why is snow rare in urban areas in the UK?

Examiner's secrets

Try to use local information. Your local council may have some data. Even the climate of the school grounds makes a useful source of data on urban climates.

Answers
Weather and Climate

Circulation systems

Checkpoints

1 Rising air produces low pressure because as air rises it expands and is less dense.
2 The Hadley cell is a convection cell that develops at the equator because of intense heating.
3 The main centre of low pressure is at the equator and the main centres of high pressure are at the poles.

Exam questions

1 The unequal heating of the earth is the main cause of the global atmospheric circulation. There is a surplus of energy at the equator and a deficit at the poles. The transfer of heat from the equator to the poles equalises the global imbalance.
2 Air at the equator is warm and therefore quite light (low density). Therefore it rises, giving a region of low pressure. At the subtropics air is colder and therefore heavier, so it sinks causing an area of high pressure.
3 A Rossby wave is the meandering pattern of the very fast jet streams.
4 A jet stream is a very fast thermal wind, located in the upper atmosphere. One jet stream is found in the subtropics and the other in temperate areas.

Weather hazards

Checkpoints

1 These areas have other benefits that outweigh the hazard problem, e.g. San Francisco is worth living in even with fog!
2 Cities produce a lot of heat and so warm air rises above them. On a cold night cold air is trapped beneath this layer. Pollution particles also encourage condensation.
3 A *tornado* is an intense low pressure system found over land. It develops as a result of the contrast between very cold air and very warm air. In addition, jet stream activity causes an intense spiralling wind system with winds of over 500 km/h. By contrast, *hurricanes* can develop as tropical low pressure systems over oceans. They are driven by warm sea waters and the spin of the Coriolis force. When they hit land they are cut off from their supply of heat (the warm waters) and so begin to die out, having first caused heavy rainfall, strong winds, flooding and mudslides in the land areas in their path.

Exam questions

1 Fog can cause large amounts of pollutants, such as SO_2 and particulates, to remain in the lower atmosphere, thereby affecting people's health. It makes driving dangerous – and aeroplanes and ferries may have to be cancelled. There is economic loss as well as a potential impact on health.
2 (a) Hurricanes develop over warm tropical seas. Intense heating causes large-scale evaporation of water. Once the low pressure system develops it is fuelled by evaporation. Winds blow inwards to the low pressure area. The whole system moves westwards. Once the hurricane reaches land it begins to die. This is because it no longer receives any latent heat from condensation of water.
   (b) Hurricanes are hazardous on account of:
      – high winds with gusts of over 120 km/h
      – heavy rain (up to 500 mm in 24 hours)
      – a rise in sea level (10 cm for every drop of 10 mb)
      – storm surges.

Air masses: Lamb's airflow types

Checkpoints

1 An air mass is a large body of air with fairly uniform temperature and humidity.
2 Warm air becomes cooler and more stable as it passes over a cold surface.
3 Cold air becomes warmer and more unstable as it passes over warm air. This instability can cause rain.

Exam question

Start with a definition of an air mass. Then state the main types of air masses to affect the British Isles. A map showing the origin (source) and type of air mass would be very useful. In each succeeding paragraph you should run through the characteristics of the main air masses to affect the British Isles.

For example, Tropical maritime air masses are stable and bring warm air from low latitudes. In winter, air is unseasonably mild and damp with temperatures of 11°C. The usual cloud type is stratus or stratocumulus with drizzle. Sea fog is common in coastal regions. In summer temperatures are typically 16–18°C. However, visibility is poor – solid particles remain near the ground.

A good answer will look at a variety of contrasting air masses. The best answers will talk about the modification of air masses, as well as the type of weather that occurs when different air masses meet to form a depression. Tropical maritime air masses commonly form the warm sector of depressions when they meet polar air masses. Similarly, if a tropical air mass moves over a cold surface, such as the cold Canary Current, it is chilled and becomes more stable (i.e. less likely to produce rain).

Characteristics of air masses

Checkpoints

1 Continental air is drier than maritime, which is moist because of its origin over the sea.
2 Tropical continental air is warm and dry because the air has come from the warm land rather than the sea.
3 Am – Arctic maritime very cold and wet
   Tc – Tropical continental warm and dry
   Ac – Arctic continental very cold and dry
   Pc – Polar continental cold and dry

Exam question

Both North America and the British Isles are in the paths of two totally different air masses in terms of temperature. At certain times of the year one type of air mass tends to dominate, e.g. in the winter it is Polar air. The exact characteristics of the air mass will depend on whether it is continental or maritime in origin. The 'battleground' scenario develops when two strong contrasting air masses meet, as this can lead to the formation of a depression (see 'Weather fronts', pages 46–47).

Weather fronts

Checkpoints

1 A front is the boundary between two air masses.
2 Depressions have low pressure because warm air is being forced to rise.
3 A warm front has a gentler incline than a cold front and brings drizzle and rain. The cold front brings heavy rain, gusty conditions and a drop in temperature.

Exam question

This is a common essay question and is divided into equal parts – describe the weather, and explain it.

The first half of the essay should describe the main weather associated with a typical depression. This should therefore take into account rainfall, temperature, cloud cover, air pressure, humidity, wind speed and wind direction. For each of these you should state what happens as the warm front approaches, at the warm front, in the warm sector, at the cold front, and behind the cold front. If you describe each of the characteristics at each of the locations you will have a very full answer.

To explain the processes is more difficult. The main factor is the difference in density between warm air and cold air. This causes warm air to rise over the cold air. Better candidates will talk about the air masses involved, their temperatures, and the intensity with which uplift takes place. Top answers will differentiate between ana fronts and kata fronts. Again, top candidates will refer to depressions as 3D phenomena in which jet stream activity at the top of the depression is responsible for removing the uplifted air and maintaining the strength of the system.

High pressure systems

Checkpoints

1 A cyclone is low pressure and brings rain. An anticyclone is high pressure and brings settled conditions.
2 In a cyclone in the northern hemisphere winds blow anticlockwise towards the centre of the low. In an anticyclone winds blow clockwise out from the centre of the high.
3 The high pressure system is caused by cold air sinking and so there are no clouds. Clear skies mean that the earth loses heat at night.

Exam questions

(a) The high is 1024 mb.
(b) The isobars are circular and spaced quite far apart. This tells us that wind speed is light or variable.
(c) With little cloud cover, night-time temperatures plummet because there is no cloud to act as a blanket and keep the heat in. Day-time temperatures are low because of the time of year (December) when the sun is low in the sky and there is little insulation. In addition, there appears to be little impact of the North Atlantic Drift (Gulf Stream), which raises winter temperatures in the British Isles. Much of the wind appears to be blowing from the north-west and north-east rather than from the south-west.
(d) (i) Problems associated with high pressure in summer include low-level pollution and poor air quality, such as smog in urban areas and low-level ozone in suburban and rural areas. Occasional thunderstorms develop after periods of high pressure.
(ii) High pressure can lead to smog, fog, freezing fog, poor air circulation and so asthma sufferers and those with breathing disorders often find times of high pressure quite stressful. Driving conditions may be dangerous, especially in fog and freezing fog.

Pollution

Checkpoints

1 Volcanic eruptions that can produce ash and gases.
2 *Sustainable* means that an ecosystem or resource should be managed so that it continues naturally and is not destroyed.

Exam question

You should use case studies to back up your argument. In general terms, pollution increases with population growth and with economic development. This is clearly shown in the model of air quality and levels of economic development. Less wealthy countries, such as Bangladesh, have poorer economies and investment in pollution control is minimal. Instead, such countries favour industrialisation and the use of cheap, inefficient energy resources, such as lignite and low-grade coal, as sources of energy. By contrast, rich countries, such as the UK, which may have gone through a process of deindustrialisation, have the capital and the technology to tackle air pollution. There is certainly more pollution in developed countries, but there are increasing levels of pollution in NICs and in developing countries. Many are related to the activities of multinational companies. However, there are notable examples of countries where economic progress has occurred without a rise in pollution – Singapore is the country that comes to mind. The insistence of its government on anti-pollution measures, such as banning the use of chewing-gum, is a case in point. It has developed economically but not at the expense of the environment.

Acid rain

Checkpoints

1 Acid rain is rain that has become more acidic because of human action. Dry deposition is particles of pollution that fall like dust.
2 SO_2 is sulphur dioxide; NO_x is nitrous oxides.
3 One that goes up by the power of 10 e.g. $10^3 = 1000$
$$10^4 = 10\ 000$$
$$10^5 = 100\ 000$$
4 If industry is accepted as the cause of acidification then there is the implication that the polluter should pay for damage or at least operate a cleaner system.

Exam questions

1 The most acidified areas are in the industrialised nations such as those in western Europe and east Asia, and the USA. However, on a local scale pollution is not in the main coal-producing and industrial areas. For example, in Europe, Scandinavia and Eastern Europe are very badly affected and it is thought that the sources of the contaminants are Britain and Germany respectively.
2 The major causes of acid rain are the sulphur dioxide and nitrogen oxides produced when fossil fuels such as coal, oil and gas are burned. These become weak sulphuric and nitric acids, sometimes with a pH of around 3. Coal-fired power stations are the major producers of sulphur dioxide, although all processes that burn coal and oil contribute. Vehicles, especially cars, are responsible for most of the nitrogen oxides in the atmosphere. Dry deposition typically occurs close to the source of emission and causes damage to buildings and structures. Wet deposition, by contrast, occurs when the acids are dissolved in precipitation, and may fall at great distances from the sources. Wet deposition has been called a 'trans-frontier' pollution, as it crosses international boundaries.

Water pollution

Checkpoints

1 The main sources of water pollution are from agriculture, industry and domestic sewage.
2 Siltation is the increase in silt going into water causing lack of clarity and, in the case of reservoirs, reducing the water storage capacity.

Exam question

The causes of freshwater pollution include:
- urbanisation, industrialisation and intensification of agriculture
- deforestation for urban growth and agriculture
- the damming of rivers
- destruction of wetlands
- mining and industrial development
- agricultural development
- increased energy consumption.

The consequences of freshwater pollution include:
- eutrophication
- acidification
- toxic contamination
- decline of water levels
- accelerated siltation
- extermination of ecosystems and biota.

An essay should describe and explain the causes and the consequences, and use case studies to back up the narrative.

The greenhouse effect

Checkpoints

1 Methane and, especially, CFCs have a much greater global warming potential than the same volume of CO_2.
2 Ultraviolet radiation breaks down ozone and oxygen into individual atoms. These atoms can then combine to form ozone (O_3).

Exam questions

1 The 'greenhouse effect' is the process by which certain gases, such as CO_2, methane and CFCs, absorb outgoing long-wave radiation from the earth, and return some of it back to earth, thereby raising the earth's temperature.
2 With increased atmospheric energy there is greater potential for more intense storms (more energy means bigger storms) as well as more frequent storms.

The monsoon

Checkpoints

1 'Monsoon' is derived from the Arabic word *mausim* meaning seasons, and refers to a seasonal wind. It is also used to describe the rains that come with the moist winds.
2 This is a breeze that blows from land to sea at night and then reverses to blow from sea to land during the day.
3 The jet stream is a high-altitude air movement from west to east.

Exam question

This essay requires a clear distinction between the winter monsoon and the summer monsoon. This is best done with the use of a climate graph and then a clear description and explanation of the seasonal differences. The main points to remember are as follows.

The monsoon is the reversal of pressure and winds, which gives rise to a marked seasonality of rainfall over north and south-east Asia. In winter, temperatures over central Asia are low, leading to high pressure. This brings outward-blowing north-easterly winds across south Asia. These dry air streams produce clear skies and sunny weather over most of India (November–May). Consequently there is very little rainfall.

By contrast, in summer an intense low pressure area develops over Asia. Strong convectional heating under clear skies also leads to low pressure drawing in warm, moist air from over the Indian Ocean. Winds are drawn into the Asian low pressure area bringing with them heavy rain. For example, Cherrapunji can receive over 10 000 mm in just four months.

The Mediterranean climate

Checkpoints

1 Winter rain is brought by westerly winds that originate over the sea and are therefore wet.
2 Rainfall decreases from west to east, with Athens receiving approximately half the amount of rainfall for Gibraltar.

Exam question

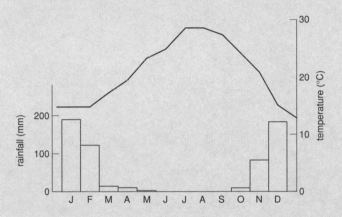

This is very similar to the monsoon question. The description is best illustrated with the help of a climate graph, while the explanation must stress the two contrasting pressure systems – high and low pressure. In summer the subtropical high pressure (STHP) dominates the area, creating weak winds and calm conditions. In winter, however, Mediterranean areas experience frontal activity and with it, associated rainfall.

Temperate climates

Checkpoints

1 Shannon is on the west coast of Ireland and is exposed to the effect of the sea. Westerly winds blow in from the Atlantic bringing rain.
2 Continentality is the effect on climate caused by distance from the sea. Continental climates tend to experience extremes of temperature.

Exam question

(a) July temperatures increase the further the station is away from the sea (Berlin highest, Shannon lowest). January temperatures are coldest further east (Berlin the coldest). This is due to the continental effect.
(b) Berlin receives heavy convection rain due to the high temperatures.
(c) Shannon is kept above freezing due to the moderating effect of the sea and the influence of the Gulf Stream.

Tropical climates

Checkpoints

1 All savanna areas have a dry season.
2 Rainforest climates show little change whereas savanna climates have a marked seasonal pattern.

3 Convectional rain is caused by a sudden upsurge of air due to intense heating. Other types – frontal (in a depression) and relief.

Exam questions

1 A descriptive account of an equatorial climate should cover:
  • rainfall total
  • rainfall seasonality
  • type of rainfall
  • temperature profile
  • seasonal range
  • daily range
  • length of growing season.
2 A descriptive account of the savanna climate should include all the checkpoints listed for an equatorial climate. Remember to use the figures – quote maxima and minima for both temperature and rainfall. The explanation needs to include reference to the effect of the overhead sun causing high temperatures which then trigger off intense convection currents and heavy rain.

Urban microclimates

Checkpoints

1 A city heat island is caused by both the structure of the city and the different air above it. The city also generates heat from traffic and buildings, which can get trapped.
2 Air is more unstable above the city and the strong convection surges can trigger thunderstorms.
3 In the rural area the clear sky brought by high pressure allows rapid cooling at night, whereas the city traps the residual heat because of the buildings.
4 The urban heat island causes precipitation to fall as rain rather than snow. If snow does fall it melts rapidly.

Exam question

This is a straightforward essay in two parts. The first part should describe succinctly the main characteristics of an urban climate, such as:
• temperature and heat island effect
• amount and type of precipitation
• wind speed and wind characteristics
• humidity
• cloud cover.
The explanation should include:
• sources of heat from industry and offices
• albedo
• specific heat capacity of buildings
• absence of vegetation and water bodies
• the impact of homes, cars and other forms of transport.
Top answers will have lots of examples, many from local studies – even from the microclimate of the school.

The study of ecosystems is a key element of geography. Within physical geography the study of the nature, structure and processes in ecosystems, together with patterns of their distribution, is the field of biogeography. Understanding ecosystems, though, is also essential in any study of the interaction of people and their environments, for we need to understand what effect people's actions may have on ecosystems. We also need to understand how ecosystems can provide resources for human use. The field of sustainable development is underpinned, therefore, by a knowledge of how ecosystems work.

Exam themes

→ The nature and distribution of the world's major biomes

→ Ecosystem structures and processes – how ecosystems work

→ Changes in ecosystems over time

→ People–environment interactions in different ecosystems

→ Forests and forest management

→ Grasslands and their management

→ Deserts and management issues

→ Tundra and polar environments and their management issues

Topic checklist

○ AS ● A2	EDEXCEL		OCR		AQA		WJEC
	A	B	A	B	A	B	
Energy flows and nutrient cycles	●	○	○				
Ecological succession	●	○	○		○	●	○
Forest and grassland ecosystems							
Tundra and polar wildernesses							
Managing desert wildernesses							

Energy flows and nutrient cycles

An **ecosystem** comprises the plants, animals and non-living components (e.g. water) that exist together in a particular location and also the processes (e.g. photosynthesis) that operate in that environment. They can be any size, from a puddle, to a lake, to one of the world's large natural vegetation zones (**biomes**), such as the Amazon tropical rainforest.

Energy flow in ecosystems ●●●

All the energy required for the processes in ecosystems is obtained from the sun. This energy is transferred through the ecosystem as follows:

→ Energy from sunlight is captured by green plants and converted into plant material and sugars by photosynthesis. The sugars are used by the plant for living processes, but some energy becomes plant material (**biomass**), an amount called the **net primary productivity (NPP)**.

→ Plants are eaten by **herbivores**, which gain their energy for life from the plant material, but also store some of the energy as biomass.

→ Herbivores are eaten by **carnivores**, who use some of the energy for living (reproduction, movement, etc.), but convert some to biomass.

→ Carnivores may be eaten by **top carnivores**.

An example of such a **food chain** is: leaves → ants → voles → hawks. In practice, most animals feed off several species, so ecosystems are really **food webs**.

Trophic levels and ecological pyramids ●●●

The figure below shows an ecological pyramid representing organisms in a pond. Each level in the pyramid is called a **trophic level**. The number of individuals and the amount of biomass decrease from level to level as there is less energy available to support life.

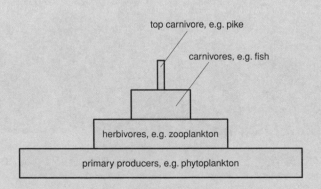

top carnivore, e.g. pike

carnivores, e.g. fish

herbivores, e.g. zooplankton

primary producers, e.g. phytoplankton

Productivity in ecosystems ●●●

The world's major ecosystems vary in productivity. High productivity is found where:

Checkpoint 1

What would be the main components of a pond ecosystem? Don't forget living (biotic) *and* non-living (abiotic) things. Try to distinguish *inputs, outputs* and *processes* in the pond ecosystem.

The jargon

Gross primary productivity (GPP) is the amount of energy photosynthesised. *Net primary productivity* (NPP) is GPP minus the energy used for respiration by the plant.

Checkpoint 2

Write a simple definition of the terms highlighted in **bold** in this section.

Action point

Draw a food web for an ecosystem you have studied.

Checkpoint 3

Try to explain why there may be only one tiger in a large area of tropical forest in northern India.

- there are high temperatures (e.g. in the tropics)
- there is a plentiful supply of water (either as precipitation or in a water ecosystem such as a lake)
- there is a plentiful supply of nutrients (e.g. where rivers carry nutrients into an estuary, or where ocean currents converge).

High NPP is found, for example, in a tropical rainforest (2000 g/m^2/year). Low NPP is found in hot deserts or tundra (e.g. 100 g/m^2/year).

Nutrient cycles

Nutrients such as carbon, nitrogen or potassium are the chemical elements that are essential to support life. Plants and animals obtain them through the food chain. Nutrient cycles show the pathways that nutrients follow within the environment. The figure below shows a simple model of the **carbon nutrient cycle**. Within the cycle:

- carbon occurs in different forms, e.g. as carbon dioxide gas in the air, as plant sugars, or as limestone rock
- large amounts of carbon are stored for long periods as rocks
- there is human interference, e.g. releasing more carbon dioxide into the air by burning fossil fuels or clearing forests.

Action point

For one of the processes within the carbon cycle, investigate from textbooks how that process operates.

Checkpoint 4

What are the main ways people interfere with the carbon cycle?

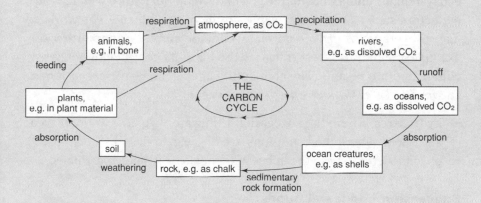

Exam questions   answers: page 84

1   The table below shows the net primary productivity (g/m^2/year) for a number of major biomes.
    (a) Explain the concept of net primary productivity. (5 mins)
    (b) With reference to specific examples, explain why NPP varies between biomes. (8 mins)

Biome	Mean net primary productivity
Tundra	140
Estuaries	2000
Temperate grassland	600
Temperate deciduous forest	1200
Tropical evergreen forest	2200

Examiner's secrets

Maximum marks require you to use named examples of ecosystems or real cases of ecosystem management issues.
Annotated diagrams can save you a lot of writing.

2   With reference to the carbon cycle, show how human activity can have an impact on natural nutrient cycles. (15 mins)

Ecological succession

All ecosystems change over time. There are small changes from year to year even in well-established ecosystems, perhaps because of slight variations in temperature or rainfall. However, such changes usually disappear as a result of negative feedback processes. Changes over long periods of time are known as **ecological succession**, and may result from natural environmental changes or from the impact of human activity.

Primary succession

Primary succession occurs when a new environment is formed. This could be due to:

→ new land being created by tectonic activity, e.g. a new volcanic island
→ other geomorphic processes creating virgin land, e.g. through the formation of sand dunes by wind.

Colonisation by plants progresses over time from pioneer species such as algae to moss, grasses, flowering plants and trees, which become established as soil forms and becomes richer with humus. Animals move in as the food supply grows. Eventually a mature, stable ecosystem, or **climax community**, will develop. Its nature will depend on factors such as rock type, water supply or the climate. The world's major vegetation zones are normally **climatic climax communities**; for example, coniferous forest (taiga) is the natural ecosystem in many cold temperate regions. The community at each stage of succession is known as a **sere**. Various types of sere occur:

→ hydroseres, where the succession is in water, e.g. in a wetland area
→ xeroseres, where the conditions are dry, e.g. in desert
→ psammoseres, where the succession occurs on sand, e.g. dunes.

Succession in coastal dunes

Sand dunes provide a good example of primary succession. The succession proceeds from bare sand through the growth of coarse succulent grass, such as marram grass, which develops a thick root mat, and low-growing herbs. This stabilises the dune and adds humus. Other species such as shrubs and small sand-tolerant trees then colonise, such as willow and birch. Grasses and weeds (e.g. fescue) then spread to create dune grassland, which is invaded by thistle, ragwort and bracken. The figure at the top of the next page shows a transect through dunes illustrating the succession.

The main pressures on dune ecosystems are from people, through trampling, pollution, inappropriate uses (e.g. cycling), and from grazing. These pressures can be managed by:

→ estate management – maintaining and fencing paths, etc.
→ habitat management – controlling scrub, or planting marram grass
→ visitor management – providing signposts, leaflets and trails.

Action point

Choose a climatic climax community, and find out the details of its plants, animals and structure.

Checkpoint 1

Write a definition of each of the *types of succession* and each of the main types of *sere*.

Action point

Take notes on a case study of coastal dune succession and management. Good examples are Studland Bay, Dorset; Oxwich Burrows, South Wales; and Les Mielles in Jersey.

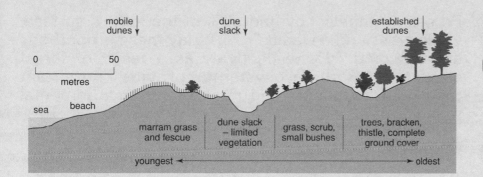

Secondary succession

Secondary succession occurs where a climax community is affected by a permanent environmental change. This could be caused by:

→ a natural change, such as climate change
→ human impact, e.g. through the clearance of woodland for farmland or the draining of a wetland.

Naturally, such a change would produce a new stable climax community. Where people maintain a sub-climax community for their own needs (e.g. farmland, or heather moorland maintained by burning) then a **plagioclimax community** develops.

The New Forest – a plagioclimax community

The New Forest, in Hampshire in southern England, is an example of a managed plagioclimax community. Its mixture of heathland, wetland and open woodland has been maintained for almost a thousand years by grazing, tree planting and drainage. Without active management it would revert to mixed deciduous woodland. Despite being the product of human management, it is now a protected 'natural' landscape.

Exam questions   answers: page 84

1 The data below show the process of colonisation and succession on the newly created volcanic island of Surtsey, Iceland.

Year	Number of moss species	Number of flowering plant species
1965	5	3
1968	8	6
1971	39	8
1974	65	15

(a) Explain the process shown by these data.
(b) With reference to specific examples, show (i) how succession occurs in a sand dune environment, and (ii) the management issues that result from this process. (15 mins)
2 'Without careful management most "natural" British plant and animal communities would not exist.' With reference to specific examples, explain this statement. (15 mins)

Forest and grassland ecosystems

Forests originally covered 60% of the earth's surface, while grassland covered 20%. Today these proportions are 30% and 40% respectively as a result of forest clearance. Both ecosystems provide important resources for people and raise many environmental management issues.

Checkpoint 1

Check the temperature and precipitation conditions under which each type of forest will be found.

Forest ecosystems ●●●

There are many different types of forest, each of which has a distinctive ecological structure. The main types are:

→ tropical broad-leaved forest (monsoon forest or tropical rainforest) is found where temperatures and rainfall are high throughout the year
→ temperate deciduous forest
→ coniferous forest, also known as taiga, which is found in middle/high latitudes and also at high altitudes elsewhere.

Action point

Choose a country as a case study of tropical forest management. Brazil, Indonesia and Malaysia are amongst the best examples.

Tropical broad-leaved forest

The clearance of the tropical broad-leaved forests has been an issue of major international concern, for some 20 million hectares per annum are lost through logging. The reasons for clearance are:

→ demand for tropical hardwoods (e.g. mahogany) from MEDCs
→ pressure to clear land for agriculture in LEDCs:
    → to meet the needs of a rapidly growing population
    → to produce export agricultural goods (e.g. beef)
→ economic pressure on LEDCs to export valuable forest products
→ for firewood at the edges of forests in densely populated regions.

The environmental effects of forest clearance include:

→ the loss of trees that play a major role in absorbing atmospheric carbon dioxide and releasing oxygen
→ a rapid loss of fertility in soils:
    → most of the nutrients in the ecosystem are located in the trees, and are lost with the trees
    → the exposed soils are leached by the heavy rainfall
→ soil erosion, because of the heavy rainfall on exposed soils
→ flooding, resulting from the erosion of soils into the rivers
→ loss of species because of the loss of habitat – tropical forests contain large numbers of species, many unknown, which might be a resource, e.g. for future pharmaceuticals.

Checkpoint 2

Reducing the loss of species is called 'maintaining biodiversity'. Why is this important to people?

The strategies used to reduce forest clearance and its impacts are:

→ decreasing demand for tropical forest products such as mahogany
→ encouraging tropical countries to export much higher-value processed timber (e.g. plywood) rather than logs
→ stricter management of logging to reduce wastage, encourage replanting and reduce erosion after clearance
→ using alternative fuel sources in LEDCs, such as biogas digesters, mini-HEP schemes, or developing fuelwood plantations.

Check the net

Identify what Greenpeace says about forest management at:
http:www.greenpeace.org

Temperate deciduous forest

Little temperate deciduous forest remains, because of clearance for farmland, e.g. in England. Major environmental issues are the management of such forests for recreation and to provide sustainable supplies of timber, and the damage caused by atmospheric pollution (acid rain).

Taiga

Taiga is cold temperate coniferous forest. Large areas remain untouched, e.g. in Siberia. Key management issues are the impact of acid rain and the use of sustainable forestry techniques.

Grassland ecosystems

There are two major types of grassland ecosystem.

Tropical grassland

These areas are known as **savannas** in Africa and **llanos** in South America. They occur where rainfall is insufficient to support forest. Key environmental issues are:

→ pressure to develop farmland because of rising populations
→ the creation of biosphere reserves (e.g. game parks)
→ pressure on soil and vegetation in game parks from increasing animal numbers and from tourism.

Strategies to deal with these issues include the culling of animals to maintain numbers within carrying capacities and the zoning of reserves to focus tourism in some areas and protect others for wildlife.

Temperate grassland

These are the **prairies** of North America, the **steppes** of Russia and the **pampas** of South America. Most have been exploited for farming, either for crops or grazing. This has resulted in problems of water supply and soil erosion (e.g. the Great American Dust Bowl of the 1930s). Strategies to reduce soil erosion include:

→ avoiding fallow periods, by leaving crop stubble or mulching
→ improving irrigation
→ providing windbreaks and shelterbelts
→ contour ploughing
→ reducing animal numbers to the land's carrying capacity.

Checkpoint 3

What are the main countries from which tourists visit the game reserves of East Africa? Think about where responsibility lies for erosion by tourists.

Test yourself

Draw a flow diagram to show the stages/steps that link the placing of animals in game reserves and the start of soil erosion.

Checkpoint 4

List three advantages and three disadvantages of biosphere reserves.

Action point

Find out the difference between *sheet erosion* and *gully erosion*.

Examiner's secrets

Use a quick sketch map to locate a case study.
Examiners want you to show that you can argue a case as well as know the facts, so you need to show that you do or do not agree with particular statements.

Exam questions   answers: page 85

1   With reference to a specific case study, discuss the environmental, social and economic issues resulting from tropical forest clearance. (12 mins)

2   Critically evaluate the management strategies used to reduce the environmental impact of game reserves in savanna regions. (12 mins)

Tundra and polar wildernesses

On the pole-ward side of the coniferous forests lie the tundra zones, which merge into the ice-covered polar regions. In the southern hemisphere the continent of Antarctica is a major wilderness region. Many regions have permanently frozen ground (permafrost) near the surface. These regions provide extreme physical challenges to people and raise many environmental issues. The main regions are:

→ the northern wilderness regions
  → northern Canada and Alaska
  → northern Scandinavia
  → northern Russia
→ Antarctica.

Tundra ecosystems

Tundra ecosystems are characterised by:

→ plants that can tolerate low temperatures, wet conditions and strong winds, such as grasses, mosses, dwarf willow and birch
→ plants that can reproduce in the short summer period
→ animals that can withstand the cold (e.g. arctic fox).

The northern wilderness regions

These regions are characterised by:

→ very low population densities and large unpopulated areas
→ some remnants of traditional cultures (e.g. the Inuit of Canada)
→ severe transport difficulties
→ harsh environmental conditions
→ an economy based on exploiting mineral resources and on some summer tourism (e.g. coastal Greenland)
→ severe social and economic deprivation amongst permanent residents.

The principal environmental issues relate to mineral exploitation. Oil extracted at Prudhoe Bay, Alaska, for example, is transported by surface pipeline to Valdez in southern Alaska, raising problems of:

→ oil pollution from pipelines, especially in earthquake zones
→ oil pollution from tanker accidents (e.g. *Exxon Valdez*, in 1989)
→ the impact of oil installations and oil towns through pollution and waste disposal issues
→ disruption of animal migration patterns, as pipelines act as barriers, e.g. for caribou
→ damage to permafrost and plant communities by vehicles, or the use of heated pipelines for utilities or oil transport.

Action point

Find out the climate data for a number of places in the tundra or polar regions.

Test yourself

Put the following species into a tundra food web diagram: arctic fox, caribou, lichen, reindeer, arctic wolf, polar bear, grass, arctic hare, moss, ptarmigan.

Checkpoint 1

What are the main transport difficulties in northern wilderness regions?

Action point

Make some detailed notes on a case study of major mineral extraction, e.g. oil at Prudhoe Bay, coal in Spitzbergen, gas in Russia.

Antarctica

●●●

Antarctica is almost entirely covered by vast ice sheets, and supports only limited animal and plant life. It has no permanent inhabitants, and is managed by 39 nations under the terms of the 1959 Antarctic Treaty. Despite being a wilderness, Antarctica has many environmental issues, e.g.:

→ pressure to exploit its mineral resources; although exploitation has been prohibited for 50 years, the continent has vast resources of coal, and the continental shelf may have large oil reserves
→ the Antarctic 'ozone hole' resulting from pollution by MEDCs
→ the impact of tourism – cruise liners and visitors cause pollution and disrupt penguin and seal colonies
→ environmental damage at Antarctic bases – pollution from the US McMurdo base has been a problem, as has the construction of runways.

The table below compares the resources of the Arctic and Antarctic regions:

Resource	Arctic	Antarctic
Minerals	Oil (Canada), gas (Russia), coal (Spitzbergen), plus many minerals	Coal and many minerals
Fish	Cod, haddock and capelin – stocks under pressure	Only two species exploited
Whales	No commercial whaling	Some scientific whaling
Environment	Wilderness resource	Wilderness resource, valuable for scientific research

Strategies for managing Antarctica depend on the continuation of the agreements of the Antarctic Treaty.

> "Any activity relating to mineral resources other than scientific research shall be prohibited for at least 50 years."
>
> Antarctic Environment Protocol, 1991

Action point

For each of the issues described in the Antarctic environment, think through what might be done to address the problem.

Checkpoint 2

Why are countries so eager to have Antarctic bases and territories?

Checkpoint 3

What differences are there between the issues facing Antarctica and those facing the Arctic?

Exam questions   answers: page 85

1   With reference to a specific case study, discuss the social, economic and environmental issues that arise from mineral exploitation in tundra or polar regions. (12 mins)

2   'Wilderness areas should be not be exploited for their economic resources.' Critically evaluate this statement and comment on its relevance to *either* northern Canada/Alaska *or* Antarctica. (15 mins)

Examiner's secrets

Group your ideas together, e.g. 'social factors'.
Including a simple sketch map to show the locations you are talking about is very helpful.

81

Managing desert wildernesses

Deserts cover 30% of the earth's land and occur in over 60 countries. Deserts, or arid zones, have less than 300 mm of rainfall per year, but semi-arid zones with 300–500 mm of rain per year have many similar management issues. The main issues in the world's hot deserts regions are:

→ water supply and management
→ soil conservation
→ fuelwood management
→ desertification.

Hot desert ecosystems

Low precipitation and high temperatures support ecosystems where:

→ animals and plants have adapted to avoid the heat (e.g. nocturnal animals) and to find and retain water (e.g. cacti store water)
→ reproduction is timed to coincide with wet periods
→ plants can tolerate very dry and very salty conditions.

Managing desert wildernesses

Water supply and management

Water supply problems arise because rainfall totals are low and often unreliable, and desert water is often saline. Stored water evaporates rapidly (30% of the water in Lake Nasser in Egypt evaporates each year). In addition, rapid population growth means demand for water increases rapidly, and traditional sources (e.g. groundwater wells) may be depleted or lost. Management approaches include:

→ rainfall harvesting, i.e. catching and storing as much rain as possible
→ increasing water extraction from groundwater
→ taking water from rivers, particularly those flowing into deserts from outside (e.g. the Aswan High Dam scheme on the River Nile)
→ desalinisation of sea water or saline lake or groundwater
→ piping water from 'water-rich' zones
→ maximising water use through trickle or point irrigation methods, and reducing evaporation from crops by mulching.

Soil conservation

Low and unreliable rainfall means that farmland may easily be damaged by salinisation, overgrazing or drying out. Desert soils are thin and lack much humus, so soil erosion by wind, or by rain when it occurs, is a major problem. Possible solutions include the maintenance of soil cover at all times by using cover crops or mulching, and managing irrigation very carefully. Tree planting may be helpful, so long as the species chosen are not highly demanding of water. Limiting numbers of pastoral animals to the land's carrying capacity is important.

Action point

Check that you know the names and locations of the world's major deserts and semi-arid regions – then sketch their location on a world outline map as a test!

Checkpoint 1

What adaptations do plants and animals make to enable them to live in hot deserts?

Action point

You will need to know a detailed case study of (a) a large multipurpose water management scheme (e.g. Aswan) and (b) a local-scale scheme. Make notes on one of each and learn them.

Checkpoint 2

What are the main strategies to prevent soil erosion?

Fuelwood

The main source of energy in deserts in LEDCs is fuelwood. Limited availability and growing populations mean that trees and brush are cleared. This increases risks of erosion and desertification. International agencies are promoting the use of alternative fuel sources, e.g. solar power, biogas digesters, etc.

Desertification

Desertification affects the semi-arid regions of the world (e.g. the Sahel of Africa). It results from:

→ natural causes – unreliable rainfall and periods of drought cause marginal land to lose vegetation
→ population growth – increasing demand for resources means that unsuitable land is cultivated or land is overgrazed
→ poor land and water management
→ problems of fuel supply and the use of fuelwood

The figure below is a model of the desertification process:

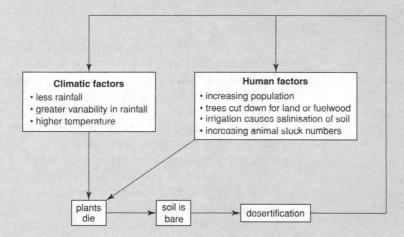

> "The cost of desertification is around $26 billion a year."
>
> Gaia Atlas of Planet Management, 1994

Checkpoint 3

What are the main alternatives to using wood for fuel in desert-margin countries?

Check the net

For case studies of the management of desertification, look at the UN Food and Agriculture Organisation web site: http://www.fao.org

Test yourself

When you have read these two pages, draw up a table listing the causes of each management issue covered, and the strategies for dealing with each one.

Exam questions   answers: pages 85–6

1 How far are the main environmental management issues in deserts the result of low and unreliable rainfall? (15 mins)

2 Study the data below, which show the population and mean annual precipitation of Tanzania's semi-arid regions:

Year	Population (millions)	Average precipitation (mm)
1948	2.25	325
1958	2.61	360
1968	3.49	215
1978	5.98	240
1988	7.01	335
1998	7.98	340

(a) Describe and explain what the data show. (7 mins)
(b) With reference to a specific case study, critically evaluate the strategies that might be used to reduce the threat and impact of desertification. (15 mins)

Examiner's secrets

The word 'critically' asks you to think of both the good and the bad features of what you are discussing.

83

Answers
Ecosystems

Energy flows and nutrient cycles

Checkpoints

1 Living components of a pond: microscopic plants and animals (e.g. amoebae), insects (e.g. pond skaters), higher plants (e.g. pondweed), higher animals (e.g. fish), frogs, decomposers (e.g. bacteria), land-based animals (e.g. heron, water rat). Distinguish between living things (a) in the water, (b) in the bottom deposits. Non-living components: water, dissolved gases (e.g. oxygen, carbon dioxide), minerals (nutrients).
Inputs: streamflow, direct precipitation, minerals from erosion, temperature.
Processes: living processes (e.g. movement, respiration, sensitivity, growth, reproduction, excretion, nutrition), chemical processes (e.g. solution), physical processes (e.g. temperature changes, transport of materials by water).
Outputs: water outflow, evaporation, living things (e.g. birds, seeds).

2 *Biomass* – the total mass of all living things or of a particular species in an ecosystem.
Net primary productivity – the amount of energy stored as plant material in an ecosystem.
Herbivores – animals that feed only off plants.
Carnivores – animals that feed only off animals.
Top carnivores – animals that feed off carnivores but are not preyed upon themselves.
Food chain – the route by which energy passes between living things by feeding.
Food web – the complex feeding inter-relationships between living things.

3 A tiger is a top carnivore. It requires a large territory (ecosystem) and biomass to support its energy needs.

4 Burning fossil fuels, quarrying limestone, clearing forest.

Exam questions

1 (a) Explain each part of the term. *Net* means it is the energy stored after life processes; *primary* that it is energy accumulated by plants. Overall it is a measure of how much energy is converted to biomass. Explain the units that NPP is measured in by saying what they mean, i.e. grams is biomass, m^2 relates it to a unit of area, and year is the measure of time over which it is measured.
  (b) Choose examples of ecosystems to represent the list. Explain why each has a high or low NPP. Show the effect of temperature, water and nutrient availability, and explain the impact of each of these.

2 Describe *and* explain the carbon cycle. Use a diagram/flow chart. For each stage explain the processes, and then show how human activity interferes, e.g. how burning fossil fuels alters flows of carbon and adds it to the atmosphere. Explain what effect these changes to flows may have, e.g. the effect of increased carbon dioxide in the atmosphere and its impact on the 'enhanced greenhouse effect'. Don't forget to use real examples that you have studied, e.g. Amazon rainforest clearance.

Ecological succession

Checkpoints

1 *Primary succession* – the development of an ecosystem from an initial state without plants and animals.
Secondary succession – the development of an ecosystem from an existing ecosystem as a result of change in environmental conditions.
Hydrosere – an ecosystem where water is the major controlling influence on its nature.
Xerosere – an ecosystem where drought is the major influence on its characteristics.
Psammosere – an ecosystem where the presence of sand is the major influence on its characteristics.

2 *Natural factors* – fire, drought, storm damage, flooding, disease.
Human factors – clearance of vegetation, introduction of a new plant or animal species, pollution.

3 A plagioclimax community is one in which the main influence on its nature and characteristics is the interference of people.

Exam questions

1 (a) Describe the increase in species, quoting actual figures, and identifying when growth was fastest. Then use key words, e.g. *primary succession, pioneer species,* to show that you understand succession. Explain *why* species increase, and the impact of increasing numbers of plants and animals on food supply and predator–prey relationships.
  (b) (i) Start by explaining the examples you will use and locate these with sketch maps. Describe the process of dune succession. Name species at each stage, and draw a cross-section model.
    (ii) Identify the management issues in your chosen examples. List them to start with, then show how each is caused and managed. Distinguish between social, economic and environmental issues.

2 Identify key words in the quote, i.e. *management, communities.* The question wants you to show that you realise most British communities are plagioclimax. Choose examples and show how they are created and managed by people. The best examples will be ones where you can use fieldwork evidence and local knowledge. Discuss and explain secondary succession, what the climax community might be, and what the impact of people is in each of the changes. Try to show how the impact of human interference can be minimised, but also that such plagioclimax communities will only be sustained by management.

Forest and grassland ecosystems

Checkpoints

1 *Tropical broad-leaved forest* – temperatures 27–33°C all year round, precipitation 1500–2500 mm per year.
Temperate deciduous forest – temperatures 0–25°C,

with growing season of 8 months or more, precipitation 500–1500 mm per year. *Taiga* – temperatures –25–20°C with a growing season of 4 months or more, precipitation 300–1500 mm per year.

2 Maintaining biodiversity is important: to protect known resources, to protect resources we may not yet know about (e.g. future medicines), to stop unknown ecological changes, to maintain the diversity of species for humans to enjoy, and for moral reasons for the protection of species.

3 Main origins of East African tourists are European countries (especially UK, Germany and France), North American countries (especially USA and Canada), and Japan. The responsibility for erosion lies directly with the tourists who cause the damage, but indirectly with the tour companies who make a profit from tourists, and the national governments who exploit their resources for tourist income.

4 *Advantages* – preserve species, enable controlled breeding programmes, enable research, provide a visitor attraction and generate income. *Disadvantages* – costly to establish, costly to maintain and protect, can be a target for poachers, may lead to overprotection and then to the need for culling, may increase erosion, and may change the natural ecosystem processes in the area.

Exam questions

1 Start by describing the scale and location of forest clearance, and outline the reasons for it in one paragraph. Then discuss the issues – group them under the headings of environmental, social, economic. Use examples, or a case study throughout. A good example or case study would be of forest clearance in Indonesia or in Brazil. Be sure to include the views of all the interested parties, e.g. LEDC government, logging companies, indigenous peoples, multinational companies, Western consumers, environmental groups, etc.

2 Show you understand the question by explaining briefly the growth and location, with reasons, for game parks. Then work through the issues (e.g. increasing animal numbers), describing the strategies used to deal with each, and their effects. Illustrate each with examples, or choose a particular case study that shows most of the key issues, e.g. game parks in Kenya or Zambia. Show the advantages and disadvantages of game parks in relation to social, economic and environmental issues.

Tundra and polar wildernesses

Checkpoints

1 The main transport problems are: huge distances; difficult terrain for road building, e.g. mountainous or permafrost areas; costs of maintenance; winter weather (grounding of aircraft, snow blocking roads and railways); cost of building transport systems.

2 Antarctica has huge potential resources. Countries want to make sure they have a 'say' in how these may be used in

future and also wish to 'have their share' if resources can be exploited.

3 *Arctic* – main issues are political, since the territories of major superpowers converge there. Fewer political issues of resources, but many environmental issues of exploitation. *Antarctica* – political issues of ownership and future use of resources. Small-scale local environmental impact issues. Need for protection.

Exam questions

1 Start by describing the scale and location of mineral exploitation in tundra/polar regions, and the reasons for it in one paragraph. Then discuss the issues, grouping them together under the three headings of environmental, social, economic. Use examples, or a single case study throughout, and be sure to locate a case study with a sketch map. A good case study is the exploitation of oil from Prudhoe Bay in Alaska.

2 Start by showing that you know what wilderness areas are by trying to provide a definition with some different sorts of examples. In particular, discuss the meaning of the word 'wilderness' – is this wilderness for people to enjoy, or is it to be protected from people? Choose one of the regions and describe the arguments for/against mineral exploitation there. You could group your ideas as social, economic and environmental issues. Discuss the views of the range of interested parties, e.g. local indigenous people, multinational companies, environmental groups, etc. Finish by saying how far you agree/disagree with the statement, and why.

Managing desert wildernesses

Checkpoints

1 *Plants* – deep roots, small leaves with waxy layers, water storage (e.g. baobab), rapid reproduction cycle when water available, seeds withstand long periods of drought before growing. *Animals* – nocturnal or burrowing animals to avoid heat; can withstand long periods without water; sense organs keep out sand (e.g. camels' long eyelashes); shaped to avoid heat (e.g. camels' large feet).

2 Main strategies to avoid soil erosion are: maintaining soil cover at all times (e.g. by mulching, keeping soils wet), planting erosion controls (e.g. tree shelterbelts), keeping animal stock levels within the soil's carrying capacity.

3 Alternatives to fuelwood include: biogas digesters, mini-HEP schemes, fuelwood plantations, solar-powered cells, and animal power.

Exam questions

1 Show that you know the main environmental features of deserts with reference to precipitation, temperatures, arid and semi-arid regions, quoting actual figures. Locate them in relation to the main desert and semi-arid regions of the world. Then go through each of the main environmental management issues (e.g. desertification, water supply, soil erosion) and show what factors cause each. Be clear

about the difference between natural and human factors in causing the problems. Use examples of each, or a case study to illustrate them all. A good case study is of desertification in the Sahel region of West Africa. Finish by drawing a conclusion about the question, i.e. answer the question.

2 (a) Describe the trends in each column and explain each one. Emphasise population growth and variability of rainfall. Talk about population growth curves (J-shaped and S-shaped curves). Discuss the rainfall reliability in semi-arid and desert margin regions. Stress that the figures show a *potential* issue, needing management to ensure that the problems that might result are minimised.

   (b) Choose a case study – use a sketch map to locate it. Good case studies include the Sahel region of West Africa. Discuss each of the strategies to deal with desertification, grouping them as soil management, water management, agricultural management, population control, etc. Explain how each works and what social, economic or environmental issues each strategy may raise. For each, conclude why it is or is not successful (i.e. be critical!).

This chapter forms the basis for about one half of your studies in geography. The nature of the main specifications does mean that topics are studied at both A2 and AS, as the chart below indicates. The emphasis is on the explanation of current patterns of population settlement and economic activity, and explaining the patterns and distributions. Because geography is about change, the modern focus is upon the nature of change and explaining the spatial changes that are occurring. Behind the economic and social factors that explain many distributions of human activity lie environmental and political factors. The latter are a major focus for geographical explanations of human activity.

Exam themes

The key themes are about population change and the pressure this puts on planning and resources; changes in settlement pattern; and economic activity within regions, countries and world-wide.

Topic checklist

O AS ● A2

	EDEXCEL A	EDEXCEL B	OCR A	OCR B	AQA A	AQA B	WJEC
Population: demography	O	●	O	O	O●	O	O●
Migration	O	●	O	O	●	●	O●
Demographic issues	O	●	O	O	O●	O●	O●
Urbanisation	O●	O	O	O●	O	O	O
Rural settlement	O	●	O	O	O●	O	O●
Internal structure of cities	O●	O	O	O	O●	O	O●
Social geography of cities	O●	O	O	O	O●	O	O
The central area and service sector	O●	O	O	O	O	O	O●
Issues in cities	O●	O	O	O	O	O	O
Leisure in cities	O	O●		O●	●		
Non-renewable resources		●	●	●	●	●	
Renewable and sustainable resources		●	●	●	●	●	
Energy			●	●			O
Agriculture			●	●			O
Agro-ecosystems and agricultural change in Europe		O	●	●		●	
Challenges for agriculture			●	●			
Can the world feed itself?	●	●	●	●	O	O	
Industrial location	●	●	●	O	O	●	O
Global shift	●	●	●	O	O	●	●
Service industries		●	●		●		
Tourism		●			O	●	O●
Newly Industrialising Countries (NICs)	●	●	●	O	O		
Industrial decline	●	●	●	O	O	●	O●
Government involvement in economic development	●	●	●	O			O
Development and disparity		●	●				O●
Trade and aid		●	●		O	●	
New global order		●	●		O	●	

Population: demography

Population geography tries to explain the location, number and changes in the people of an area over time. It is a dynamic study because people are born, die and move. We measure population in censuses, which are inevitably always out of date.

Distribution and density ●●●

→ 75% of the world's population live on 5% of the earth's surface, which is 13% of the land area.
→ 66% of people live within 500 km of an ocean.
→ All large concentrations are in the northern hemisphere between 10° and 55°N, with the exception of parts of South East Asia.

The figure below shows the components of population change.

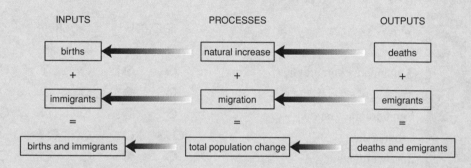

We can map different distributions of peoples at different scales. This is often used to show distributions of ethnic minorities in cities.

The jargon

Population density is the total number of people divided by the total land area.

Demographic information ●●●

The figure below shows the demographic transition model.

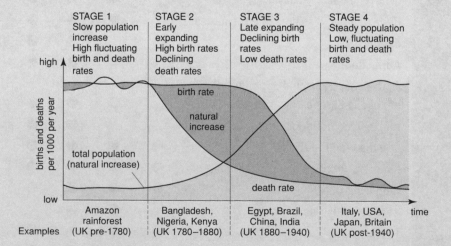

Action point

Can you draw sketch maps to show the distribution and density of population in a country that you have studied?

Checkpoint 1

What other ways are there to show population distribution?

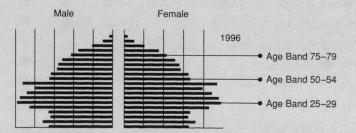

Population changes due to the interaction of births and deaths

	Population Density per km^2	Crude death rate per thousand	Crude birth rate per thousand
Monaco	31 000	N/D	N/D
Macau	21 560	N/D	N/D
Hong Kong	5506	N/D	N/D
Singapore	4650	6	16
Bangladesh	800	12	36
India	274	10	29
Sri Lanka	269	6	21
UK	**238**	**11**	**14**
Germany	228	12	10
Italy	189	10	10
Nigeria	114	16	45
France	104	10	13
Uganda	83	19	52
Ghana	69	12	42
Kenya	48	12	45
Burkino Faso	35	18	47
South Africa	32	9	31
USA	28	9	16
Zimbabwe	27	12	39
Brazil	18	8	25
Russia	9	12	11
Canada	3	8	15
Australia	2	7	15

Exam questions

answers: page 142

1   Using of examples from a range of countries, show the characteristics of countries at different stages of the demographic transition model.
(15 mins)

2   Describe the characteristics of the population pyramids shown below. Suggest a type of settlement or country which each one might represent.
(10 mins)

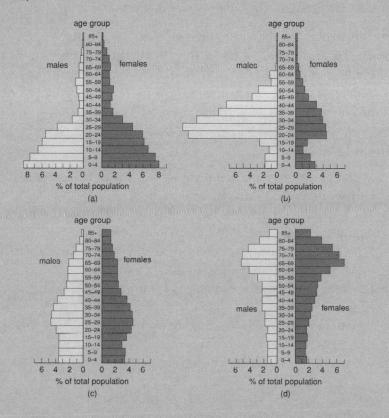

Examiner's secrets

Make sure that you can define precisely the various types of demographic data such as fertility rate, life expectancy, crude death rate.

Checkpoint 2

Look at the table. Where are the sparsely peopled regions and why is their population density so low?

Examiner's secrets

Always have a bank of quickly drawn population pyramids to illustrate answers about population structure.

Migration

Migration frequently emphasises natural population changes when people move to areas of natural growth and away from areas of natural decline. **Migration** is the movement to take up residence, whereas **circulation** describes our daily commuting, movement for holidays, and nomadism. Migration is the cumulative result of many individual decisions.

Examiner's secrets

Make sure you have examples of all the types of migration in the diagram.

Types of migration ●●●

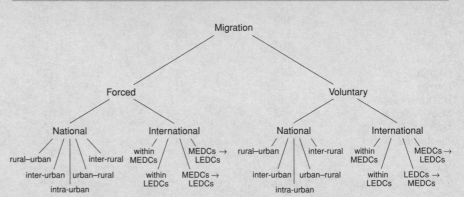

Checkpoint 1

Why are people migrating to the south-east of the UK?

Why people migrate ●●●

→ Push–pull factors
→ Harsh environments
→ Marriage
→ Economic advancement
→ Persecution and political factors

→ Quality of life
→ Population pressure
→ Forced
→ Labour migration

Why and how do people migrate?

In 1885, Ravenstein wrote his laws of migration:

→ The greater the distance, the fewer people migrate.
→ The bigger the pull, whether it be in numbers or wealth, the greater the flow of people.
→ Some people migrate in stages – intervening opportunities delay the full move.

Examiner's secrets

The examiner will look for the names of people associated with theories. For instance, Lee's model is about intervening opportunities.

The migrants

→ Migrants are often male and young, e.g. those who migrated from the Caribbean to the UK in the 1950s.
→ The less skilled migrate in search of opportunities, e.g. the Irish migration to the USA in the nineteenth century.
→ Skilled workers also migrate in search of opportunities, e.g. bankers moving to Singapore in the twentieth century.
→ People migrate from tyranny, e.g. from Hitler's Germany, or from Kosovo in 1999.
→ People migrate to new planned opportunities, e.g. into the Brazilian heartland.

Migration at a national scale ●●●

→ In the nineteenth century, people in the UK moved to the coalfields from rural poverty in search of work: the process of urbanisation. This is still a feature of cities in the LEDCs.

→ In the early twentieth century there was a movement towards London and the south-east.

→ Since the 1930s people have moved to the outer edges of cities: the process of suburbanisation.

→ Since the 1960s there has been a net movement of people away from urban areas back to small towns and villages: the process of counterurbanisation.

→ Other forms of migration have taken place, e.g. retirement migration to the south coast and to south-west England.

→ There is some migration back into cities – reurbanisation – mainly by young, high-flying, single people and the very rich.

What are the effects of migration? ●●●

The exporting area:

→ loses its breadwinners

→ becomes more elderly
→ loses its gender balance

→ loses services
→ suffers from land being abandoned

The receiving area has:

→ pressure on services and utilities such as water supply
→ demands for housing
→ a younger and possibly gender-imbalanced population structure
→ overcrowding
→ more labour availability

Migration in the USA can be illustrated like this (excluding Alaska and Hawaii):

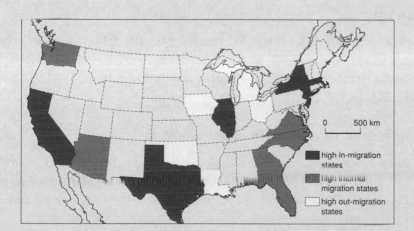

0   500 km

■ high in-migration states

▨ high internal migration states

□ high out-migration states

Exam questions   answers: page 142

1 Attempt to justify a classification of migration. (10 mins)

2 With reference to a range of examples, examine the role of government policies in managing population growth and migration. (15 mins)

Action point

Do you have case studies of international migration and its causes and effects, e.g. nineteenth-century Irish to the USA, South Asians to the Middle East, Turks to Germany?

Checkpoint 2

Explain the areas of in-migration and out-migration in the USA.

Demographic issues

Project tip

If you live in or near an area with a higher than average elderly population, e.g. Bognor, Budleigh Salterton, Worthing, see if you can identify the impact that this has on the provision and location of local services.

Links

See Development and disparity, pages 136–37, and Health and welfare, pages 151–62.

The jargon

When there is exponential growth of a population this is called a *population explosion*.

Checkpoint 1

In what way are the elderly a different issue in LEDCs?

Checkpoint 2

Why are numbers of the elderly growing so rapidly in Western Europe?

Demographic issues relate to many other aspects of human geography and especially to development and health and welfare. The most important issues are population control, ageing and life expectancy. Society's attitude to gender is also an important field of demographic study. Migrants and refugees raise issues such as the concentration of immigrant groups in particular areas. Out-migration also affects an area.

Overpopulation and population control

LEDCs are especially prone to population explosions. This is due to **life expectancy** rising, birth rate not falling and health control improving. Some countries are also trying to increase population.

Types of policy
Expansionist policies

→ Restrict availability of birth control, e.g. Ireland and Ghana
→ Make abortion illegal, e.g. Guernsey
→ Give family allowances to those with more children, e.g. France
→ Support for working mothers, e.g. Eastern Europe until 1989
→ Improve health care
→ Restrict roles for women in society, e.g. the Taliban in Afghanistan

Control policies

→ Promote birth control, e.g. Singapore
→ Permit abortion
→ Encourage sterilisation, e.g. India
→ Control the number of children in a family, e.g. China (1980s)
→ Run governmental advertising campaigns, e.g. India

Ageing

In 1993 16% of the UK population was over 65. By 2020 this is expected to rise to 19% and in 2050 (when many of you will be retired) to 29%. The elderly are dominated by females with a F/M ratio of 234:100 among those over 80.

Distribution of rising life expectancy

→ Retirement migration to the south-west and south coast of England
→ Out-migration of younger age groups from central Wales
→ Older people are left behind in inner cities because of low income and by out-migration of the young
→ In all areas, numbers swelled by the 1940s baby boom are now reaching retirement

Effects on policies

→ Restriction of conversion of properties to old people's homes, e.g. Bournemouth

- → Diversification of local economies to counteract immigration of old dominates any change
- → Companies set up to provide homes for the elderly
- → Where the elderly vote is large, this is likely to be reflected in policies, e.g. law and order – neighbourhood watch
- → Health provision pressures, ranging from care to geriatric beds
- → Policies to employ the fit elderly, e.g. some supermarkets have begun such policies
- → Increased services for the elderly, such as care and nursing homes, and access to retailing geared to the elderly

Gender

Women's roles in society have demographic economic and social causes and effects, including attitudes to work and child rearing, attitudes to female children, and attitudes to migration. There can be imbalances in gender roles, e.g. in the north-east of England where there has been traditionally less work for women because the main employment was coal mining and shipbuilding. In contrast, Lancashire textiles had a tradition of female work.

Issues resulting from migration

The concentration of ethnic groups in Greater London is shown in the map below.

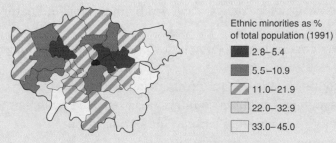

Ethnic minorities as %
of total population (1991)
- 2.8–5.4
- 5.5–10.9
- 11.0–21.9
- 22.0–32.9
- 33.0–45.0

- → Many immigrant areas have a skewed population pyramid.
- → Facilities and services are geared to the dominant group, e.g. mosques, temples, specialist retailing and travel agencies.
- → An influx of immigrants can cause other groups to leave the area.
- → Education provision may have to cater for young people whose first language is not that of the host society.
- → Gender imbalances result from labour migration.
- → The spread of diseases such as AIDS is partly brought about by migration. This increases the number of orphans and decreases the working population, especially in LEDCs.
- → Refugees may bring similar issues to the host area.

Exam questions answers: page 142

1 Discuss the impact of the changing structure of the population on the provision of services in a named country. (15 mins)

2 What might explain the pattern of ethnic population shown in the map above? (10 mins)

Action point

Make some notes on the different attitudes of societies in MEDCs and LEDCs to women and how this affects changes in population numbers, and roles in the economy and in society.

Project tip

Issues to do with ethnic migration and refugees need very sensitive treatment and are probably best left alone.

Link

See ghettos, page 101.

Checkpoint 3

Make a list of the issues that an influx of refugees might raise. What is done to prevent the issues from causing conflict?

Examiner's tip

If a question says a 'named country' make sure that you name the country. Note that Africa is *not* a country, and using such an example will not gain you marks.

Urbanisation

Of the world's population, 43% live in cities and by 2005 more people will live in cities than in rural areas. **Urbanisation** is the term we give to the process of becoming urban. It is part of a model illustrating the processes of population movement in and around a city.

The cycle of urbanisation

➜ People leave the countryside for the cities, and population also grows naturally in the city.
➜ People get better jobs and move to better areas in the city.
➜ People leave the city for the towns and villages beyond the city.
➜ People yearn for city life and return to live close to the heart of the city.

Urbanisation is very rapid in percentage terms. In Europe the process is associated with the Industrial Revolution. It involved a rapid expansion of the urban area, with people working in secondary jobs. Urbanisation is a social process because it changes a rural society into a more stratified one, where people are wage-earners and live separately from the owners of the new factories. Urbanisation in LEDCs is happening in the twenty-first century and involves both a pull by the city and a push from the countryside.

➜ Rapid demographic change is caused by migration and natural increase.
➜ Socio-economic change is due to a shift from subsistence agriculture to a market economy.
➜ Behavioural change due to the changing way of life.
➜ Under-employment in the countryside is replaced by unemployment in the city – urbanisation without industrialisation.
➜ Urbanisation leads to primate cities.
➜ It also leads to shanty towns/favelas/bustees.
➜ Shanty towns, favelas, bustees.

A table of urban percentages by regions is shown below, for 1960–2020.

Table 3.1 Percentage of urban population, by regions, 1960–2020

	1960	1980	1990	2000	2020
World total	34.2	39.6	42.6	46.6	57.4
Developed regions	60.5	70.2	72.5	74.4	77.2
North America	69.9	73.9	74.2	74.9	76.7
Europe incl. CIS	56.9	67.7	70.9	73.4	76.7
Oceania	66.3	71.5	70.8	71.4	75.1
Developing regions	22.2	29.2	33.6	39.3	53.1
Africa	18.8	27.0	32.5	39.0	52.2
Asia	21.5	26.6	29.9	35.0	49.3
Latin America	49.3	65.4	72.0	76.8	83.0

Source: *United Nations, Urban and Rural Population Projections 1950–2025: The 1985 Assessment* (New York, 1986)

Action point

Place the following terms against the points opposite: *reurbanisation, urbanisation, counterurbanisation, suburbanisation.*

"Urbanisation is a dynamic force in society whereby values and properties recognised as being urban are propagated."

Carr 1987

Checkpoint

Can you define *a primate city*? Name one.

Action point

Can you draw a sketch map showing the Shanty towns in a city of your choice?

Project tip

Projects about the issues surrounding the use of a brownfield or greenfield site are manageable and can score highly if they are well researched.

Action point

Can you list the reasons for the development of shanty towns and provide an example of them in, e.g. Latin America? What are governments doing to help to eliminate them?

In the late twentieth century, cities in MEDCs began experiencing a process called **counterurbanisation**.

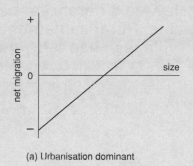

(a) Urbanisation dominant

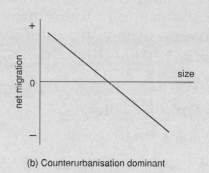

(b) Counterurbanisation dominant

Why did these changes take place in the late twentieth century?

→ Jobs in more pleasant rural areas – consultancies
→ New towns and planned growth of small towns put more jobs within easy reach of the country
→ Improvements in rail and road transport
→ Branch plant growth
→ Diseconomies of scale in cities
→ Improvements in communication – teleworking from home through the internet
→ Retirement migration to the countryside
→ Planning policies such as greenbelts making areas desirable

Reurbanisation is bringing people and jobs back to cities. The jobs are in new activities, e.g. Canary Wharf financial district employing young, affluent professionals who then live in Docklands. Much movement is onto brownfield sites in areas that have deindustrialised. It is associated with urban renewal.

Exam questions   answers: page 143

1   What factors explain the distribution of the major cities shown in the table below? (15 mins)

2   List some of the similarities and contrasts between urbanisation in MEDCs and in LEDCs. (15 mins)

City	Country	1990
Tokyo	Japan	25.0 m
New York	USA	16.1 m
Mexico City	Mexico	15.1 m
São Paulo	Brazil	14.8 m
Shanghai	china	13.4 m
Mumbai (Bombay)	india	12.2 m
Los Angeles	USA	11.5 m
Beijing	china	10.9 m

City	Country	1990
Calcutta	india	10.8 m
Buenos Aires	argentina	10.5 m
Seoul	Rep. of S.Korea	10.4 m
Osaka	Japan	10.3 m
Rio de Janeiro	brazil	9.6 m
Paris	France	9.3 m
Tianjin	china	9.1 m

Cities in LEDCs shown in *italics* and shaded grey.

Rural settlement

Action point

Does this type of village exist today? In what ways do people try to retain this idyll?

Link

Counterurbanisation, page 95.

Rural settlements were once those associated with people living and working in the countryside. Because of changes in the agricultural economy, and counter-urbanisation, there are now fewer differences between urban and rural settlements. Rural settlements are smaller than towns, e.g. villages, hamlets and isolated farms.

Form of villages

Villages may be sited at:

→ a water supply, spring lines
→ a route focus
→ access to suitable land
→ a defensive site.

Villages can be linear or nucleated or dispersed. Nucleated villages are found at a natural focus, a junction, change of transport or defensive point. Some contain greens. Many have grown owing to infilling of empty sites and by accretion around the edges.

Dispersed settlement has come about because of:

→ inheritance – dividing land between all the sons (e.g. gavelkind)
→ effects of Parliamentary Enclosure – farms amalgamated onto one area of land
→ effects of *remembrement* in France and Germany 1950–80 when dispersed field farm holdings were unified
→ effects of planned settlement policy – Ijsselmeer polders
→ government policy to reorganise marshlands in Italy – Pontine Marshes.

Measuring dispersion

It is possible to attempt to measure dispersion by using a technique called **nearest neighbour analysis**. The nearest neighbour statistic ranges from 0.0 = clustered, through 0.23 = linear clustering, 1.0 = random, to 2.15 = regular pattern.

Commuter villages

Counterurbanisation gave rise to commuter villages. What have been the consequences?

→ New social classes based on service sector work in villages.
→ Reduction in availability of homes for those who work and live in rural areas.
→ Public housing and the lower-paid are clustered in some villages.

→ Fewer low-order goods available because people shop in superstores.

→ Services for the more affluent, e.g. antique shops.

→ Decline in public transport.

→ People can live in rural areas because there is better access to jobs in more distant locations.

Second homes ●●●

Some properties are acquired as second homes as a result of rural depopulation and farm abandonment, e.g. in Tuscany. Changes in agricultural practice also make properties available, e.g. for barn conversions. The rise in second homes results in improved upkeep of the settlement and employment for those who maintain and improve property. It also causes rural house prices to rise which can provoke opposition. The age structure is likely to alter, and the settlement may be 'dead' out of season.

Rural depopulation ●●●

This is a form of population migration. The causes are:

→ loss of jobs in agriculture

→ mechanisation of agriculture

→ poor housing conditions in rural areas

→ land degradation due to overcropping and drought

→ perceived lack of provision of rural services

→ marriage leading to a move to a partner's home

→ labour migration to the towns.

The consequences are:

→ abandoned property and fallow land

→ more second homes

→ loss of services and public transport

→ loss of social cohesion in the rural community

→ ageing population because the young migrate

→ more travelling to find goods and services

→ loss of male labour, especially in LEDCs.

Project tip

Good topics for projects include: investigations into changing form and function of a village; a contrast between a village within a greenbelt and one beyond a greenbelt; migration into a village – does it conform to ideas about either migration or counterurbanisation; using censuses to look at depopulation of a village; a study of second homes in an area and their effect on the area.

Examiner's secrets

Do not get to caught up with the sensational. For instance, when discussing the drawbacks to second homes don't just focus on the burning of second homes. Mention the issue but spend more time on the more important issues such as affordable homes for rural workers.

Checkpoint 1

Why do people leave the countryside in MEDCs?

Checkpoint 2

Why do people leave the countryside in LEDCs?

Exam questions   answers: page 143

1   Examine the effects of counterurbanisation on the economic and social geography of rural settlements. (15 mins)

2   What are the differences between rural depopulation in MEDCs and in LEDCs? (10 mins)

Internal structure of cities

The jargon

A *model* is a representation of the real world. It attempts to show the general and not the unique. It is a way by which a geographer can test what is observed or measured against an approximate generality. The task is to explain why reality and generality either fit or are a mismatch.

Examiner's secrets

The Burgess model is now dated and is only a starting point. Use more recent models at AS and A-level. Burgess is for GCSE!

Action point

Convert these pictograms into sketch plans and complete the keys by giving them the correct numbers.

Every city is unique, but it is possible to create models that generalise how a city's land uses and social groups are arranged. Since Burgess (not a geographer but a sociologist) first modelled Chicago in 1925, others have developed better models for the structure of cities. Most of the models are culturally determined.

Hoyt's sector model, 1939 ●●●

This model was based on rent data and took account of the influence of transport routes on economic activity and on the distribution of socio-economic groups. It was based on the USA and is shown below.

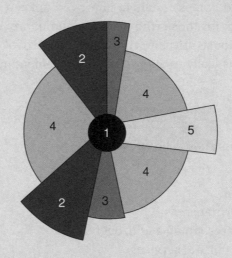

1. central business district
2. transportation and industry
3. low-class residential
4. middle-class residential
5. high-class residential

Harris and Ullman's multiple nuclei model, 1945 ●●●

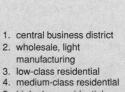

1. central business district
2. wholesale, light manufacturing
3. low-class residential
4. medium-class residentlal
5. high-class residential
6. heavy manufacturing
7. outlying business district
8. residential suburb
9. industrial suburb

This model is more realistic because it recognised the complexities of the mid-twentieth city. Nuclei developed around earlier foci such as villages. Each developed a set of economic activities and/or social distinctiveness. This theory is again from North America.

Mann's model of the British city, 1965 ●●●

This combines the earlier models. It is a more accurate reflection of growth and land use in the UK, although it is also dated.

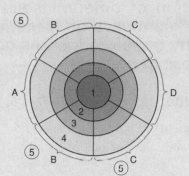

A middle-class sector
B lower-middle-class sector
C working-class sector and main municipal housing areas)
D industry and lowest working-class areas

1 city zone
2 transitional zone
3 zone of small terraced houses in sectors C and D; larger by-law houses in sector B; large old houses in sector A
4 post-1918 residential areas, with post-1945 development mainly on the periphery
5 commuting-distance 'village'

Bid rent theory ●●●

→ Rent or land value declines with distance from the city centre.
→ The values placed on each area by each land use decline at a different rate.
→ So certain land uses dominate the bidding in different areas – the rent pattern is also affected by:
  → transport
  → local government policies
  → planning

to give a complex pattern.

Examiner's secrets

Make sure that you are able to draw or sketch simple diagrams relatively quickly. Practise them. Also make sure that you can relate real places to the models. At least be able to draw a simple sketch map of the structure of your home city or a city that you have studied.

Action point

Draw a diagram to show how the land value varies from the city centre along a transect to the rural-urban fringe.

Exam questions answers: page 143

1 Outline the structure of a city known to you. Which model best reflects your description? (20 mins)

2 The figure below shows a model of the variations in the quality of life in three groups of cities.

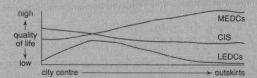

(a) State *two* criteria used to measure quality of life. (3 mins)
(b) Describe the differences shown in the quality of life for the three groups of cities. (15 mins)
(c) Choose a city which is known to you and describe how well it fits with the model of variations in quality of life. (12 mins)

Examiner's secrets

Many geography examinations ask questions that draw together parts of the syllabus. Quality of life is not a theme here but is to do with social and economic geography.

The jargon

Quality of life is based on personal satisfaction, happiness and security. It is environmentally related through housing, work opportunities, access to goods and services, pollution and leisure.

Social geography of cities

The jargon

Socio-economic status is based on family background, education, values, income and occupation.

Social geography studies different ways by which social groups are recognised, such as socio-economic groups, the behaviour of different social groups in for example buying a house, social processes such as **segregation**, and the distribution of different social groups. Housing is one of the main uses of space in cities. Much of social geography therefore focuses on the issues concerned with access to housing.

Murdie's model of social space

This attempts to link the three major elements of social area analysis, namely the ethnic, family and socio-economic dimensions of **social space** to **physical space**.

Examiner's secrets

Social geography is full of jargon. Make sure you know what the terms mean. If you use a term, **define** it.

Housing

Housing choice is only there for those who have the money to make a choice. Choice is based on (i) life-cycle, (ii) social class and (iii) life changes. Therefore there is a pattern of choice for people in the UK over time.

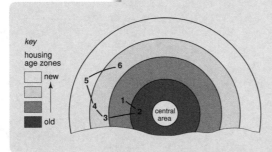

key
housing age zones
new
↑
old

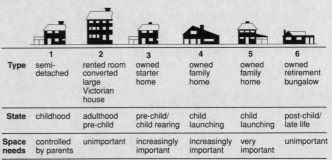

Type	1 semi-detached	2 rented room converted large Victorian house	3 owned starter home	4 owned family home	5 owned family home	6 owned retirement bungalow
State	childhood	adulthood pre-child	pre-child/ child rearing	child launching	child launching	post-child/ late life
Space needs	controlled by parents	unimportant	increasingly important	increasingly important	very important	unimportant

Checkpoint 1

List some of the social and environmental problems which result from the downward filtering of housing in the inner city.

Housing is filtered downwards through social groups (see below). This results in low-income, inner city residents (the underclass) trapped in an area with social problems. In some areas the process is reversed as professional people move into an area, and this is called **gentrification**.

Choice constraints – these are negative externalities

→ Limitations may depend on the ability to pay.
→ There are markets for different types.
→ There are fashions that increase desirability.
→ There are gatekeepers who approve finances for housing (for example, certain areas may be 'red-lined' – declared unsuitable for loans).

Action point

State whether each of the following is a positive or negative externality:

→ high crime rate
→ sea view
→ poor schools
→ a park
→ boarded-up shops
→ a hill
→ within 5 km of motorway junction
→ near a station.

Write why you have made the choice. Can some be both?

Encouraging choice – these are positive externalities

→ Areas are 'green-lined' – approved due to their future potential or being the 'right address'.
→ Some areas become gentrified.

Ghettos as segregation

Why do people cluster?

→ Avoidance – focus on the community and its religious and cultural needs
→ Preservation of the culture in home and neighbourhood
→ Defence to help new migrants and give security
→ Attack because together when threatened – riots

How do ghettos grow?

→ Spillover – gradual outward spread
→ Leapfrogging to new areas
→ Response to local policies – urban renovation pushing people out

Housing in LEDCs

Spontaneous settlements or **squatter settlements** or **informal settlements** grow rapidly from migration and natural increase. According to the UN, in 1990 squatter settlements were 33% of housing in São Paulo, 85% in Addis Abbaba and 50% in Lusaka, Bogota, Ankara, Luanda and Dar es Salaam.

How can governments help?

→ Provide a strong economy that is able to support house-building
→ Build new towns
→ Provide **site and service schemes**

Turner sees these settlements as part of a cycle: **bridge headers** become **consolidators** and hope to be **status seekers**.

Exam questions   answers: pages 143–4

1   Study the figure below, showing population growth in São Paulo, Brazil from 1870 to 2000.

(a) Describe the pattern of population growth for São Paulo.

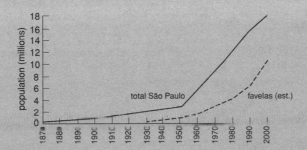

(b) What does the graph show us about numbers and the percentage of people living in the shanty towns (favelas)?

(c) Outline two possible benefits of the shanty town system.

(d) Examine some of the environmental problems associated with the rapid growth of cities in LEDCs. Illustrate your answer from a range of cities. (20 mins)

2   Discuss the factors that lead to the development and spread of areas housing cultural or ethnic minorities. (15 mins)

The central area and service sector

Action point

Can you draw your own sketch map to show the different zones in the central area/CBD of your own home town/city? Can you explain why zones do or do not exist?

In 2001 most of the jobs in the UK and in our cities are in the service sector. These jobs include working in offices, retailing, education and the public services and leisure. The central business district (CBD) is at the heart of the city. However, in the past 25 years the CBD has ceased to be the sole location for many of these activities.

The CBD ●●●

The larger the city, the more distinct each part of the CBD becomes, so it is really a central area with a retail zone, an office district and even a legal quarter, a university campus and a leisure zone.

How do geographers identify the CBD?

→ Mapping land uses – shops, offices, leisure
→ Height index – business floors as a proportion of floors
→ Intensity index – if over 50% of space is CBD
→ Council tax per metre of street frontage
→ Barriers to spread – a park, a railway, a hill, a river
→ A historic quarter that is conserved and adapted, e.g. solicitors' offices
→ Pedestrianised and/or traffic restrictions

Project tip

There are plenty of topics available in the central area involving the mapping of land uses and the explanation of the patterns to be found. Changes in the city centre are a further interesting topic.

Beyond the CBD there is the core–frame or zone in transition (see below).

Checkpoint 1

The frame contains a zone of assimilation and a zone of discard. Can you see evidence of these in a city that you have studied?

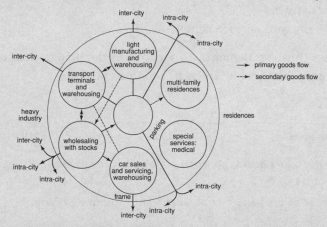

Offices ●●●

The locations of offices in West European cities are shown below.

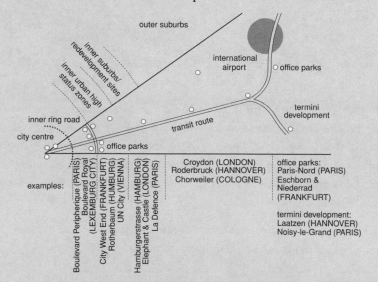

Retailing ●●●

The figure below shows a model of retailing in cities.

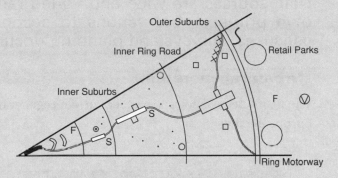

Key
⌒ City centre Pedestrianised & Traditional – Kingston Guildford High Street Cascades, Portsmouth, Arndale, Manchester, Eton Square Newcastle
⌢ Malls
▭ Suburban Centres, Leytonstone High Road
⊙ Brownfield Redeveloped Sites – Gateshead Metro Centre
• Corner Shops – mainly inner city
□ Convenience Shops/Petrol Retailing
○ Early Retail Parks – Brent Cross

○ Retail Parks – Blue Waters, Merry Hill
Ⓥ Village/Small Town – Antiques – Arundel
ſ Supermarkets & Superstores – larger the further out
××× Automobile and Fast Food Rows
══ Roads
══ Motorways
F Factory Outlets – Gunwharf, Portsmouth; Whiteley, Hants & Clark's, Somerset

Recent changes in retailing are:

→ specialist shops returning to central area – Tesco Metro
→ supermarket flight to superstores
→ arcades and covered city centre malls with specialist outlets – mobile phones, bookstores, cybercafés
→ superstores and retail parks on periphery
→ motor rows
→ factory outlets.

Both offices and retailing are **decentralising**.

Action points

Make your own notes explaining why a particular out-of-town retail development known to you was located and what were the effects on retailing elsewhere.
Do you have examples of decentralisation? Can you think of a market that has left the city core?

Checkpoint 1

For any one of the retailing types listed here, suggest reasons for its development. What might be the case raised by objectors to such a development?

Examiner's secrets

Always make sure that you have a case study city in an MEDC and in an LEDC from which you can give examples.

Exam questions answers: page 144

1 (a) Describe the built environment in the central business district of a large city in an MEDC (such as New York) in terms of:
(i) the age of the buildings
(ii) the nature of the activities that might make use of the buildings. (10 mins)
(b) How might the built environment be different to that described in part (a) in a large city in an LEDC? (5 mins)

2 New York and Singapore are major centres in the modern world economy. Outline the factors that lead to the offices of banks and other service sector firms being concentrated:
(i) in the heart of cities
(ii) in cities such as New York and Singapore. (10 mins)

Issues in cities

Action point

Can you draw your own sketch map to show the different zones in the central area/CBD of your own home town/city? Can you explain why zones do or do not exist?

The jargon

Urban regeneration is the term created by 1999 Lord Rogers' *Urban Task Force,* for a wider process of environmental and social improvements.
Urban redevelopment means starting again on brownfield sites.
Urban renewal means working within the existing built form.

The city is the focus of most of our lives in the twenty-first century. It provides geographers with a range of themes to be studied and topics to be investigated. Data sources are wide and varied ranging from your own primary observations to secondary data from censuses, surveys and planning documents.

Urban regeneration

The figure below shows a model of urban regeneration in the heart of European cities.

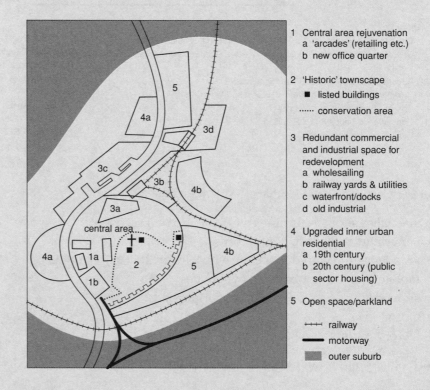

1 Central area rejuvenation
   a 'arcades' (retailing etc.)
   b new office quarter

2 'Historic' townscape
   ■ listed buildings
   ······ conservation area

3 Redundant commercial and industrial space for redevelopment
   a wholesailing
   b railway yards & utilities
   c waterfront/docks
   d old industrial

4 Upgraded inner urban residential
   a 19th century
   b 20th century (public sector housing)

5 Open space/parkland

   ┼┼┼ railway
   ─── motorway
   ▓ outer suburb

Project tip

All these initiatives could be the subject of a good local project.

Delivering Regeneration, *The Patchwork Quilt, the Audit Commission 1989,* made the following comments:

→ City Centre Partnership Companies – Coventry 1998
→ Enterprise Zones, 1981–96
→ 12 Urban Development Corporations, 1981–98 – LDDC
→ Urban Development Grants, 1982–88 – financial support for developments within the private sector
→ City Grant and English Partnerships to fund private sector in priority urban areas; merged with Commission for New Towns
→ Single Regeneration Budget (SRB) Challenge Fund 1994 – resources for up to seven years for economic, social and physical regeneration (fifth round 1999)
→ Housing Action trusts 1992 – redevelopment of social housing
→ Recycle buildings – hotels in old chapels, church to temple
→ Better use of redundant floors, e.g. above shops

Other central area issues

Derelict retailing; conversion of offices to gentrified apartments; homelessness; cars and public transport; new tram routes and environmental impacts; park-and-ride schemes.

Issues in suburbia and beyond

→ Retail parks and superstores – impact on environment and on other retailing. Office parks and science parks have an impact on employment, commuting patterns and other areas of city.

→ Leisure centres and new parks on reclaimed land, e.g. former gravel pits. Could include golf courses, new stadia and cinemas.

→ All these might be on the edge of greenbelts – raises questions about nature of greenbelts.

→ Marinas on reclaimed land or in old docks or just on coast.

→ Loss of school playing fields – not so much an issue in twenty-first century.

→ By-passes – building, noise, new edge for built-up area.

→ Sale of council houses and impact on environment.

New settlements

New towns – hardly new now as many are over 50 years old. However, there are issues to do with further growth and spread of this generally successful means of accommodating urban growth, e.g. Harlow, Essex.

Private new towns – many now developed, e.g. Bradley Stoke near Bristol or Hampton near Peterborough.

New villages and major **village expansions**, many of which are to comply with demands for housing in the affluent south-east of England.

Political factors

In most cases involving change in the city, political factors loom large. Decisions may be made to accord with political ideology, in response to pressure groups, to satisfy a forceful, rich developer or a big corporation.

In all cases involving change it is useful to ask, 'Who gains what, where, how and why?' and conversely, 'Who loses what, where, how and why?' These questions are the substance of '**Welfare Geography**'.

Examiner's secrets

Always make sure that you have detailed information on a case study city in a MEDC and in an LEDC. Your home town is a starting place but make sure that you have at least one other to use to make your answers different. Manchester, Glasgow and Leeds are three good study cities. (London is not an easy city to use – use just a segment of it.)

Checkpoint 1

Why does the country need more houses if population numbers are hardly growing?

The jargon

Ideology is a system of ideas and beliefs that a particular group holds. It is used to show the ideas and beliefs of a particular social, cultural or political group.

Exam question   answer: page 144

Consider an inner city area in the UK before and after regeneration.

(a) Describe the ways in which housing, streets and the general environment have been improved.

(b) Outline some ways in which you could use primary fieldwork and secondary research to study improvements resulting from regeneration in an urban area. (20 mins)

Leisure in cities

The jargon

Leisure refers to those activities that take place outside of the working day/week. The choice of activity is decided by the person concerned.

Recreation is leisure taking place from home. It can be active, e.g. sports and clubbing, or passive, e.g. visiting a cinema. It can be organised, such as a visit to a soccer match or a theatre, or informal, such as sitting in a park. It can be resource-based, e.g. a city park, or user-based, e.g. a dry ski slope.

Tourism is a commercial form of leisure that usually takes place away from the home base. It normally involves staying away for more than one night.

> "There will be a boom in creative learning facilitated by the Internet. We will spend more on our psychological well-being and physical appearance – leisure surgery."
>
> Richard Scase

Checkpoint 1

Can you think of how these predictions for 2010 are already resulting in patterns of urban leisure?

Leisure time is mainly used for recreation although even when we are tourists we often make use of recreational facilities. Even people at work can make use of leisure facilities, e.g. conference delegates going out for dinner or to a club. Many facilities for recreation and tourism are found in the leisure zones of the central area of a city. Hotel locations are another key leisure feature of cities. Leisure is decentralising as sports stadia move to the outskirts, e.g. San Siro, Milan, and cinema complexes move to the outer suburbs. Parks and greenbelt land should be seen as leisure areas in cities. Some people classify shopping as a leisure activity!

The elements of leisure in cities

Primary elements

Place-based	Area-based
Cultural facilities	*Physical characteristics*
Theatre	Historic streets
Concert halls	Buildings of note
Cinemas	Monuments and statues
Exhibition centres	Churches, mosques and cathedrals
Museums and art galleries	Parks and open spaces
	Water – canal and river fronts
Sports facilities	Harbours and marinas
Indoor arenas	
Stadia	*Social attractions*
Golf courses and	Liveliness – atmosphere
playing fields	Language
Amusement	Local customs and folklore
Casinos	Security
Night clubs	Suits age group
Festivals	

Education and self-improvement

Secondary elements		Other elements
Hotels	Shopping	Accessibility, information including
Restaurants	Markets	parking, signposting, maps

What has caused the growth in leisure?

→ Less important today is shorter working time – in fact working hours for professional persons have increased
→ More money to spend
→ Improved transport technology – access to out-of-town cinemas
→ Increased holiday allowances – more money for city breaks
→ Importance of well-being – parks, playing fields, gyms
→ Increasing domestication of leisure – pubs that look like home
→ Technology, from videos to stadium rock and the Internet
→ Growth of the new service class doing routine jobs in homes, so releasing time for leisure

→ Marketing of venues
→ Improved information via media

Hotels as leisure points ●●●

The figure below shows a model of the distribution of hotels and other accommodation in a city. Use your knowledge of your home town/city to see if it conforms to the model. What other leisure activities do the hotels have? What activities nearby are served by the hotels? (Think of the hotels around Disneyland Paris, for example.)

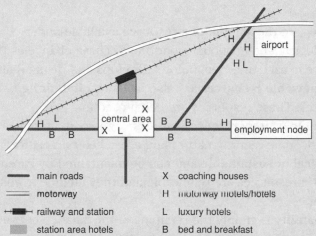

—— main roads	X coaching houses
══ motorway	H motorway motels/hotels
▬▬ railway and station	L luxury hotels
▨ station area hotels	B bed and breakfast

Greenbelts as leisure space ●●●

Greenbelts were originally established in 1938 in order to check the spread of urban areas such as London, to prevent towns merging (as in the Ruhr), to preserve the special character of towns (such as Bath), and to provide for recreational needs (e.g. Box Hill, Surrey). Some greenbelts have been used for playing fields; these may be privately owned and not available to everyone.

There is increased pressure for recreation caused by easier transport to city fringes. For example, abandoned mineral workings are now often used for recreational purposes such as sailing or retail parks. Population and development has leapfrogged the greenbelt and increased demand for recreational space from both sides. There is now also an increase in the number of golf courses on set-aside land.

Decentralising leisure ●●●

Some leisure facilities are moving out from the city centre: cinema complexes, new sports stadia and concert venues, exhibition centres, hotels.

Exam questions
answers: pages 144–5

1 Examine the factors that have assisted the growth of urban leisure. (15 mins)

2 With reference to one named urban area, discuss the impact of increased leisure facilities on the environment. (10 mins)

3 In what ways may tourist facilities also be recreational facilities? What planning problems arise from catering for both recreation and tourism in the same venues? (20 mins)

Project tip

Many of the leisure functions listed here could be the subject of a project, particularly if they are new and have had an impact on people and their environment.

Checkpoint 2

Why do hotels decentralise?

Project tip

Issues of over development in greenbelts are a good project topic. They will not only be about recreation.

Examiner's secrets

Collect useful facts for essay introductions:

→ In 1956 we spent 4% of income on leisure. In 1968 it had risen to 9%, with 2% on holidays. In 1999 we spent 17% on leisure and 6% was on holidays.

→ The richest 10% spend 20% of their income on leisure whereas the poorest 10% spend 12%.

Non-renewable resources

Checkpoint 1

Which of the following are critical or non-critical: soil, water, solar energy, forest?

Examiner's secrets

Check that you know what 'ubiquitous' and 'localised' mean. Sometimes these words are used in geography examinations.

Link

See global development, pages 136–41.

The way in which we make use of the earth's declining resources is a key theme in geography. We all use resources to live and we must appreciate that their use is a major challenge facing society.

Key concepts

→ **Stocks** are the total resource discovered and undiscovered on the planet.
→ **Resources** are the items that can be used for food, shelter or to gain a livelihood.
→ **Reserves** are the amount of a resource available under current technological and economic conditions. These change as the resource is used, new technology develops and demand alters.
→ **Non-renewable resources** are also called **capital**, **finite** or **stock resources**. These resources are being used up.
→ **Renewable** or **flow resources** are those that can be re-used or renewed. Some recur in nature and are called **non-critical**. Others are **critical** or **sustainable** and can be maintained by careful use.
→ **Aesthetic resources** are renewable and they include townscapes and landscapes.
→ **Sustainability** is simply the retaining of the balance between the availability and use of resources. **Sustainable development** is a basic idea lying behind many studies of global development.
→ Resources can be **ubiquitous** or **localised**.

Cycle of exploitation

1. Exploration and discovery, reserves large. → 2. Exploitation of major finds, reserves increase. → 3. Exploitation of minor finds, reserves declining. → 4. Exhaustion approaches, recycling and substitution.

Finite resources

1. Copper
There are 560 million tonnes of copper reserves – this will last 65 years. Some 10.7 million tonnes are mined each year (Chile 2.2 m tonnes, USA 1.8 m tonnes and Canada 0.6 m tonnes). Copper is used for electrical cables, wire and pipes, in wood preservatives and in alloys such as brass and bronze.

2. Lead
There are 120 million tonnes of lead reserves. Some 5.7 m tonnes are produced every year (Australia 0.46 m tonnes, Canada 0.35 m tonnes, USA 0.23 m tonnes, China 0.20 m tonnes). Lead is used as a protection against X-rays, and in batteries, glass and paints. Lead is also a major pollutant and killer that has caused people to look for a replacement for lead in products such as paint on toys, the replacement of lead water pipes (especially in Scotland), and lead in petrol, which is being eliminated. This will lower the demand for lead.

3. Coal

Reserves of coal could last for between 200 years (anthracite) and 400 years (lignite). It accounts for 30% of the world's energy production and its use is growing in the developing world. China uses 25% of the world's coal and its use is growing by 4%. Therefore, there is no real concern about coal supplies. (See Fossil fuels, page 112.)

Responses to declining resources

→ **Replace** production from countries in decline, e.g. Chilean and Russian copper are replacing Zambian and Japanese supplies; Australian lead production has also risen. Minerals from many new areas have a lower ore content.
→ **Recycle**, e.g. copper from cables and lead from batteries. The energy needed to recycle copper is 10% of that needed to refine it.
→ **Substitute**, e.g. lithium ion batteries, and plastic pipes substituted for copper.
→ Impose legal measures to enforce recycling, e.g. Germany 1991.
→ Impose taxes to enforce recycling, e.g. in Colombia.
→ Rio Declaration on Environment and Development 1992 – Principle 8: Eliminate unsustainable patterns of production.

Recycling

Why recycle?

→ Concern for environment – dumping waste (see Lead, above)
→ Concern for environment – energy for new production
→ Employs many people – 30,000 in aluminium recycling in USA
→ Cuts demand for raw materials
→ Can be traded – global aluminium recycling worth £375 m
→ Other case studies of recycling that you might use: aluminium, paper, plastics

Checkpoint 2

What are the costs and benefits of paper recycling?

Project tip

Local attitudes to waste disposal, and especially to the burying of waste, is a worthwhile topic that can link physical and human geography.

Exam questions   answers: page 145

1   Draw a diagram to explain how one of the following mineral deposits is formed: tin, lead, oil, coal , gravel. (8 mins)

2   What does a company consider before deciding to exploit a mineral resource? (7 mins)

3   What kind of fieldwork would be needed to investigate the impact of a mine or a quarry on an area? (10 mins)

Examiner's secrets

When asked to provide a located example, state clearly *where* or *what* – otherwise you halve your marks!

Renewable and sustainable resources

Renewable resources are those which on our time-scale are almost inexhaustible. Some people refer to *potentially* renewable resources, which take time to renew themselves. It is this group that can be depleted so that our use of the resource is not sustainable.

Renewable resources

Renewable resources are solar energy, wind, tides and water, both for power and consumption. Potentially renewable resources are air, fresh drinking water, soil, forests and animals. Forests have always been cleared: Europe has lost 50%, tropical LEDCs 6%. However, forests are increasing in Europe but declining in Africa and Latin America. European forests are 90% productive timber, tropical forests are only 50% productive timber. Some 5 million hectares are being felled in the tropics per year, which is 10–20 times the area being replanted. Only 5 African and 4 Asian countries have felling programmes.

What are the uses of forest products?

→ Construction, e.g. Japan uses 78% of timber from South East Asia, much for plywood to encase reinforced concrete
→ Paper (mainly coniferous forests), furniture and housing
→ Fuelwood for use by the local population (see page 112)
→ Clearance for plantations and cash crops, e.g. rubber and oil palm in Malaysia
→ Tourists, e.g. Nepal where camping tourists use more timber than the Nepalese.

NB The main users are MEDCs.

Policies to maintain forests:

→ replanting on a 1:1 basis or more, for every felled tree
→ reforesting marginal land not used for agriculture
→ improving genetic quality of trees
→ methods to optimise tree growth – coppicing did this in the past, giving a **sustainable yield**.

Sustainability of tropical rainforest is desirable because:

→ it might help stabilise global climates by reducing greenhouse gases, but vast replanting will be needed
→ it reduces soil erosion, silting, flash flooding
→ it retains a gene bank for future use
→ local people still have a livelihood in the forest
→ exports of forest products are maintained to benefit the country
→ the world will benefit as the forest product is traded.

Threats to potentially renewable resources

1. Urbanisation on best land. 2. Poor soil management. 3. Salinisation. 4. Wetland destruction. 5. Deforestation. 6. Overgrazing. 7. Pollution. 8. Groundwater depletion. 9. Elimination of habitats.

Resource conservation ●●●

This is the careful use and protection of resources such as soil, water, forests and fish, or coal, tin, copper and even sand and gravel deposits. As a concept it preceded sustainability and was less persuasive because it paid less attention to the fact that resources will be used.

Sustainability ●●●

To reach a stage where resources are managed in a sustainable fashion, and use is at the optimum level, Hawken has suggested:

→ businesses that are environmentally irresponsible, exploiting without care, should not be allowed to operate
→ public/international bodies are needed to protect, e.g. fish and natural aesthetic resources such as coral reefs and game reserves
→ when selling a natural resource, build in the cost of removing it from its ecosystem, in order to support renewal
→ subsidise sustainable practices such as replanting, and tax those that are not, e.g. clearing for plantations
→ reduce waste of energy, water and minerals
→ build environmental considerations into trade and loan agreements
→ slow global population growth. (Agenda 21 notes 'sustainability' in Principles 1 and 8 and implies it in many others.)

"... process of change in which the exploitation of resources, the direction of investments, the orientation of technological development and institutional change are all in harmony and enhance both current and future potential to meet human needs and aspirations."

The World Commission on Environment and Development Definition of Sustainable Development

Checkpoint 2

What are Principles 1 and 8 of Agenda 21?

Project tip

Projects in this area are difficult but the following could be investigated:

→ Aesthetic resources such as heathland and woodland and their value.
→ Reclamation of areas where a resource has been exhausted, e.g. brick clay pits, water use and re-use.

Exam questions   answers: pages 145–6

1 How is water supply managed to satisfy demands in an area? (10 mins)

2 What is the relative importance of the various energy resources at a global scale? (5 mins)

3 What is the case for and against the greater development of renewable energy in the UK? (10 mins)

4 This extract is about strategies to reduce forest destruction. 'The world possesses nearly two billion ha of tropical forest. 56 countries containing the most seriously affected forests contain 50% of the world's population. The aid agencies think-tanks produced a Tropical Forest Action Plan. The plan stresses the needs of 200 million forest dwellers and the need to assist the environment movement in poor countries. Friends of the Earth state that the plan hardly recognised the poor and displaced in the forests and there was no recognition of the fact that national debt was forcing exploitation of forests. The channels for forest conservation are: international loans for conservation, loans to countries who put forests in order, a body to supervise a green forest product label, and rich countries leasing forest to exploit and replant in LEDCs in exchange for debt cancellation after 1997.' *The Independent.*
(a) Can you extract the factors causing forest destruction? (b) Provide a matrix or table identifying the advantages and disadvantages of the strategies being suggested in this extract. (12 mins)

Examiner's secrets

Sustainability is a key geographical concept in the 21st century.

Energy

Energy is essential to human life. It provides heat, light and power and is the life blood of society. Energy resources are getting scarcer and people are having to conserve energy and discover new sources so that future generations will have a sustainable future. There is a politics of energy resulting from our efforts to obtain energy which may involve local conflict or, eventually, a global conflict.

Energy types ●●●

Primary energy includes all basic sources, e.g. coal, oil, solar, wind. **Secondary energy** is produced from the primary source, e.g. electricity, petrol.

There are four main types of energy: **mechanical** – clockwork; **electromagnetic** – electrical energy; **chemical** – photosynthesis; **nuclear** – radioactive fission. Energy flows through an economic system as a natural resource. It is a raw material for industry, agriculture and transport, and provides heat and light.

Fossil fuels ●●●

Fossil fuels form about 75% of global energy use (more in MEDCs).
Oil 30% of primary energy consumption in UK. Use falling in MEDCs and rising in LEDCs. 54% used in transport, 20% in homes and 19% in industry. Production dominated by the Middle East, so the politics of production should be understood.
Coal 30% of primary energy consumption in UK. Use in UK has been affected by politics, e.g. miners' strike. Large reserves – UK reserves could last 1000 years. World reserves mainly concentrated in USA, Russia and China.
Natural gas 25% of UK primary energy. Production dominated by North America, Europe, Russia. Reserves are twice those of oil and found in many LEDCs and NICs. Some producers are now searching for substitutes, e.g. shale and coal-seam methane in the USA.
Nuclear power 9% of UK primary energy. 90% generated in MEDCs. Future use controversial because high cost of building/ decommissioning, disposal of spent fuel, impact of disasters Chernobyl.
Hydro-electricity 7% of primary energy in UK. Mainly produced in MEDCs although some notable dam schemes provide power for LEDCs, e.g. Kariba, Akosombo and Aswan dams.

Biomass ●●●

Biomass accounts for 6% of global energy consumption, of which wood is the main source. Some 21 countries depend on wood for 75% of their energy (it is 90% in Mali and Burkino Faso). Animal dung can be up to 90% of energy consumption in Indian villages. Straw is also a source.

Fuelwood gathering takes 10% of Peruvian women's time; in Kenya women spend 24 hours a week collecting firewood.

Renewable energy

→ Solar power needs high sunshine totals, e.g. deserts, Pyrenees
→ Modern wind generation needs strong winds, e.g. Palm Springs, California
→ Wave power needs a coastal region: still under development
→ Tidal barrages need a large tidal range, e.g. Rance estuary, Brittany
→ Geothermal only where there are igneous rocks, e.g. Larderello, Italy and New Zealand

These all have an environmental impact.

Biomass – energy from plant and animal waste

Solar power

Tidal and wave power

Wind power

Geothermal power from heat below ground

Hydro-electric power (HEP)

Checkpoint 2

What are the effects on the local society of time spent by women on gathering fuelwood?

The politics of energy

Conventional electricity generation needs water for steam and for cooling, so river and coastal sites are most common. Local conflicts could be caused by the size and height of the chimney stack (visual intrusion), the global environmental threat of acid rain, loss of land and valuable ecosystems, e.g. marshes, and disruption in construction including moving turbines to the site.

Nuclear power stations cause conflict because disasters have alerted people to the dangers on an international scale; the cost of decommissioning is also now known – £230 million for one.

What other arguments are used? Hydro-electric dams fare little better in MEDCs: dams spoil aesthetic beauty in mountains (tourists do not want to see dams – or do they?), and can destroy wetland ecosystems (e.g. Hainburg on Danube not built); conserve energy rather than build more power stations; in LEDCs the loan costs can cripple an economy (e.g. Itaipu on Parana River has saddled Brazil with debt); lakes gradually silt up and salinisation can occur.

Examiner's secrets

Do not confuse steam that you see coming from cooling towers with smoke pollution and acid rain. Pollution in the form of acid rain comes from the power station's chimney.

Action point

At the global scale consider: the role of OPEC, the impact of OPEC on exploration, the Gulf War.
At a national scale consider: UK energy policies since 1979, basic energy for development, survival in LEDCs.

Checkpoint 3

Why can large dams cause a rise in a country's debt?

Exam questions   answers: page 146

1   What is the global energy crisis? (8 mins)

2   How is the uneven distribution of energy resources affecting the economic development and energy policy of one LEDC and one MEDC? (20 mins)

3   'The viability of renewable energy resources depends on physical geography and not the ability to develop them.' Discuss this statement. (15 mins)

Examiner's secrets

Make sure you know a case study of the environmental effects of a dam, e.g. the Aswan dam, on river flow, irrigation, and the Nile delta. What are the benefits and drawbacks of the scheme? The Colorado is another good example.

Agriculture

There are two basic types of agriculture: **subsistence** and **commercial**, which distinguish food for consumption and food for sale. It is also possible to classify according to the intensity of production: **intensive** and **extensive**. In the past, food production was limited by transport and markets, as Von Thünen theorised nearly 200 years ago. The physical environment, farm size and tenure also controlled patterns. Today, governments, bio-technologies and global demand together with ever-rising populations have brought about major changes in agriculture.

Agriculture

Types of agriculture in the world.

Arable	Crops
Pastoral	Animals
Mixed	Crops and animals
Intensive	High input /high yield
Extensive	Low input /low output
Subsistence	Produce for own use
Commercial	Produce for sale

Action point

For each of the types of agriculture you should be able to explain where it is located, what are the characteristics of the system, what is grown or reared, what are the effects on the environment and what changes are taking place. Finally you should say why the changes are taking place and what their effect is on people, the economy and the environment. Fear not! You only need to have one LEDC type and one MEDC type – and maybe one other just in case.

Checkpoint 1

Why is subsistence agriculture under threat?

Examiner's secrets

Always have a bank of diagrams to use in answers. Practise the diagrams of the Von Thünen model and Sinclair's modification as they are key knowledge on land use around a market.

Commercial agriculture

At a national scale it can be intensive, with high-input levels of working capital (machinery, seeds, fertilisers and livestock), fixed capital (the land and buildings), technology (control of light, hydroponics), labour (although less so today), fertilisers and pesticides (genetically modified?) and a high output per hectare of land.

Most intensive are: market gardening, horticulture, glasshouse crops and pick-your-own farms. Freezing and canning has led to intensive specialised cropping regions, e.g. around pea-freezing plants. Even crop growing and grazing are more intensive thanks to the removal of hedges and all the inputs listed above. Plantations may be another form of intensive agriculture.

Irrigation agriculture is mainly practised in warmer regions but still occurs in the UK – not very efficient because up to 60% of the water is not used by the plants! The drip-feed method is more efficient but is capital intensive.

What is grown where?

Three groups of factors inform a farmer's decisions.

1 Physical limitations of the area: climate, soil and relief.
2 Social limitations imposed by the culture that determine how land is held (tenure) and how land is organised in terms of size.
3 Economic limitations such as the costs of seeds and fertilisers, access to a market for the produce and the demand for the produce shown by the price that can be obtained.

Overarching points 2 and 3 is the role of government.

Government and agriculture

Until the mid-1980s, the UK government aimed to keep us fed, keep productivity high, keep prices fair and enable farmers to have a good livelihood. Helped by the European Community, CAP prices were supported to provide farmers with income, and farmers of marginal land were assisted. Other activities were encouraged in rural areas. Since the mid-1980s production has risen too fast so that surpluses have resulted. Price support is costly. Inputs such as fertilisers and modern machinery are damaging the environment. Therefore policies have shifted to:

→ controlling production by **quotas** – milk 1984
→ **guarantee thresholds**
→ **set-aside** – farmers paid to take 20% out of production
→ **farm diversification** – new crops such as oilseed rape. Threats to the environment are starting to be addressed
→ **ESAs** (Environmentally Sensitive Areas)
→ Nitrate Sensitive Areas.

Threats to agriculture

→ **Soil erosion** results from overgrazing, ploughing, stubble burning; soil **poaching** – compaction by heavy machinery leading to erosion
→ **Salinisation** – salt accumulation caused by poor irrigation practice
→ **Hedgerow removal** – destruction of natural field ecosystems
→ Changing landscape is a threat to an aesthetic resource for tourism
→ Tourism and recreational use of land – trampling, crop circle makers
→ In LEDCs – overcropping, population pressures, drought, pests, disease, war

Checkpoint 2

How does the government influence agriculture?

Project tips

You could use old photographs, maps, press reports and pressure group reports to look at the effects of agriculture on landscape change.
Try finding out what has been done to set-aside land.
Do pick-your-own farms affect the environment?

Exam questions   answers: page 146

1 How have farmers modified rural environments? (10 mins)

2 Using examples from farming in LEDCs, examine the effects of drought, disease and pests on agriculture. (10 mins)

3 Why are so many people affected by starvation in Africa? (15 mins)

Agro-ecosystems and agricultural change in Europe

Agriculture has evolved from controlling the local ecosystem in order to produce food, to the simple farming systems of today. The food web is simple and the trophic levels have been reduced. Agricultural change in Europe does illustrate both the environmental, economic and social effects of policies on the landscape and people in the countryside.

The jargon

Scientists talk about the *gene revolution* as well as the earlier *Green Revolution*.

Link

Genetic modification, page 119.

Characteristics of agro-ecosystems

Crops and animal species are fewer. They are more uniform because of genetic modification by breeders to ensure more yield or faster-growing beasts. Genetic modification results in less genetic diversity (e.g. cloned sheep). Energy flow is concentrated into a few levels.

In a grazing ecosystem, biomass is larger due to fertilisers. Herbivore mass is larger due to food additives and the optimum volume of plant biomass eaten. Once killed, the stored energy and nutrients are removed from the system. Less dead matter and humus is returned to the system. The system then depends on artificial inputs of fertilisers and farmyard manure.

Soil depends on chemical fertilisers. Their overuse leads to erosion, salinisation, waterlogging and desertification. In some areas there is very high energy consumption, especially oil for fertilisers and transport: 17% energy used in USA is by agriculture – 4% on crops, 2% on livestock, 6% on processing food and 5% on distributing food – resulting in the cheapest food in the world.

Pesticides

They were developed to control insects, weeds, fungi, roundworm and rodents.

The dangers include:

→ crops and animals develop genetic resistance
→ they kill the natural predators of the pests
→ they can result in new pests
→ more pesticides are developed and the cycle continues
→ pesticides are blown on the wind and affect natural areas or drain into rivers, killing fish
→ there is a threat to human health – sheep dips, nitrates in drinking water
→ they become concentrated up the food chain – the figure below illustrates the concentration of DDT up the food chain.

Checkpoint 1

What is a *food chain*?

> "10 units of non-renewable fossil fuel are needed to put 1 unit of food production on the table in the USA whereas with traditional intensive farming 1 unit of energy puts 10 units of food energy on the table."
>
> Tyler Miller

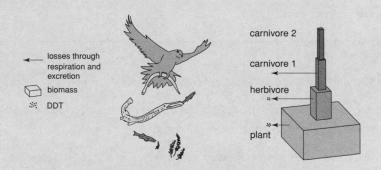

losses through respiration and excretion

biomass

DDT

carnivore 2

carnivore 1

herbivore

plant

There are natural solutions to pests, e.g. using their natural predators or rotating the crops.

The Common Agricultural Policy

The European Union aims to be self-sufficient. Some 70% of EU funds go to agriculture. Agricultural land is 42% of land area and it employs an average of 6% in the EU (3% UK and 21% Greece). The policy was reformed in 1992 because of the high cost of subsidising over-production and selling off production on world markets with further subsidies. The changes were:

→ reduce overproduction by using less intensive methods
→ switch arable land to grassland
→ restructure farmers' incomes so they are not reliant on price support and high production
→ protect the natural environment, e.g. reduce chemical applications
→ develop the natural potential of the countryside by preserving areas that maintain high-quality ecosystems and biodiversity, e.g. hedgerows
→ continue encouraging technology and genetic engineering to obtain higher yields and more productive strains but . . .
→ lower fertiliser inputs and lower yields required
→ remove land from productive use for long-term set-aside of, for instance, wetlands
→ set up afforestation projects
→ support conversion to organic farming.

Organic farming

Organic methods use a mixture of traditional techniques, e.g. free-range chickens with new ecological methods such as modern seeds. It takes 2–3 years for the effects of non-organic farming, e.g. chemical fertilisers, to leave an area. Labour costs are higher, e.g. weeding, and there is a less intensive use of land. EU standards came into force in 1993. Organic farming might have some negative effects such as soil compaction by machinery, and it is less efficient to work in smaller fields.

The move to organic farming came about due to scares over long-term effects of genetically modified foods and fertilisers on human health – salmonella in eggs, BSE from contaminated beef, sewage in feedstuffs. Products can be more than 20% more expensive, with perhaps higher profits for retailers.

The area given to organic farming in the EU quadrupled in 1987–93, encouraged by government subsidies to convert, although only 2–3% of food sold in MEDCs in 2000 was organic. In the USA, some 24% of shoppers buy organic products regularly.

Action point

Draw diagrams to illustrate the structure of an ecosystem and energy and nutrient flows in an ecosystem. Check page 74. Draw similar diagrams for an agro-ecosystem.

Examiner's secrets

Do not see everything as 'gloom and doom'. While modern agriculture may have its faults, it has enabled the world to be better fed. Always retain a balanced view.

Checkpoint 2

Why does it take time to switch to organic farming?

Project tip

It might be possible to contrast a farm using organic methods with one that doesn't (probably best for someone with farming contacts).

Exam questions   answers: pages 146–7

1 What are the consequences for the environment of modern agricultural production in MEDCs? (10 mins)

2 Outline some of the effects of the Common Agricultural Policy on agriculture and rural areas. (8 mins)

3 Contrast organic farming with traditional modern farming in terms of inputs, methods and outputs. (15 mins)

Challenges for agriculture

Much of our geographical study is about the application of our knowledge and theories to issues that arise in the environment. The Green Revolution was one response to the challenge of overpopulation that threatened parts of the world. Genetically modified food is a more recent challenge caused by the ever-present demand for cheaper food.

Feeding the world

There is a cycle of poverty in LEDCs – malnutrition, decreased resistance to disease, high infant death rate, decreased ability to work and learn, shortened life expectancy leading to poverty and further malnutrition/famine.

Solutions: 1. Cultivate more land, which involves destroying forests or irrigating dry land; 2. Use science to enhance output; 3. Eliminate waste; 4. Food aid from MEDCs; 5. More trade of surpluses; 6. Forgo cash crops for food crops – causes balance of payments problems.

The Green Revolution

→ Launched by the FAO in 1963.
→ Involved developing HYV cereals, e.g. IR8 rice, to help feed regions of high population growth such as Asia, the Philippines and parts of Latin America. IR8 was developed at the International Rice Research Institute (IRRI) in the Philippines and maize and wheat at the International Maize and Wheat Improvement Centre in Mexico (IMWIC).
→ High inputs of fertiliser, pesticide and water to obtain high yields.
→ Increasing intensity and frequency of cropping (two harvests a year).
→ New dwarf varieties came after the 1960s, which were faster growing, allowing three crops of rice a year in parts of the tropics.

Outcomes

→ There were increases in grain production of over 70% between 1960 and 1990. Yields rose by over 3% per annum. India doubled its rice yield in 15 years.
→ Fertiliser use is increased.
→ The Revolution focused on cereals; other food crops in LEDCs did not have increased yields, so other food crops rose in price.
→ Fossil fuel consumption increases for inorganic fertilisers and fuel.
→ Yields rise but fall back due to soil erosion, salinisation, pollution and waterlogging.
→ Countries are able to feed their growing population.
→ The cost of buying fertilisers is high, and poorer farmers and those with small landholdings are unable to afford them, so they do not use them.
→ Less labour is needed, which encourages rural depopulation and migration to cities.
→ Former suppliers of grain no longer have export markets.

→ Capital input needs, e.g. tractors, have increased the dependency of LEDCs rather than diminished it as was anticipated.

→ Global food supplies doubled 1960–80 but only grew by 20% 1980–90 because HYV limits were reached.

→ Population growth has increased the demand for grains.

The table below shows the countries that benefited most from the Green Revolution.

Country	Change in rice area		Change in rice yield	
	million ha	percentage	metric tonnes/ha	percentage
China	3.1	10	1.76	57
India	4.0	11	9.66	51
Indonesia	1.6	21	1.72	83
Bangladesh	1.4	15	0.44	27

Genetically modified (GM) foods ●●●

The 'Gene Revolution' is the successor to the Green Revolution. Why GM food?

→ Increases yields and decreases fertiliser usage.

→ More resistant to disease, so less pesticide use.

→ More regular supply of products for food-processing industries.

→ Higher profits for food producers and processors.

→ Gives foods a longer shelf-life and reduces waste.

By 1995, there had been 2500 field tests in the USA. In the UK field tests commenced in 1998. Effects and outcomes:

→ Interbreeding between GM and non-GM crop may have effects as yet unknown.

→ There are human health worries – not proven but not ruled out.

→ Increased simplification of agricultural ecosystems.

→ The risk of super-organisms in the environment.

→ Reduced global biodiversity.

→ There have been benefits in terms of drugs and the production of vaccines, e.g. GM alfalfa used for cholera vaccine.

A social cost-benefit analysis is being used to assess and evaluate the impact of the benefits to society against the risks of genetic modification

The jargon

This topic is full of jargon. You might find these on web sites:

→ *Biotechnology*: the use of biological processes
→ *Genetic engineering*: the transfer of genes from one organism to another
→ *Genetically modified organism (GMO)*: any plant or animal that has been genetically engineered
→ *Novel food*: used by companies to describe genetically engineered food
→ *Pharming*: medicinal products from genetically engineered plants and animals
→ *Transgenic*: refers to animals that contain genes from another species
→ *Desirability quotient*: societal benefits/societal risks
→ *Greatly desirable quotient*: large societal benefits/small societal risks
→ *Uncertain desirability quotient*: potentially large benefits/ potentially large risks
→ *Small desirability quotient*: large societal benefits/much larger societal risks

Action point

Which of the quotients listed above applies best to GM foods? Use the list on the right to decide. You could even try to suggest other economic changes that might fit the other quotients. Who should decide: the biotechnology companies who are often parts of huge corporations (TNCs); governments; or environmental pressure groups?

Exam questions   answers: page 147

1   Outline some of the effects of the Green Revolution on agriculture in a selected LEDC. (12 mins)

2   Describe the effects of drought on the economic and social life of one LEDC. (12 mins)

Project tip

This topic is best avoided for projects unless you are also studying biology. Even then it should be avoided.

Can the world feed itself?

The theories of Malthus, Boserup and the Club of Rome are essential for the study of population geography and also the study of the ability of the world to feed itself.

Malthus's theory, 1798

1 Population grows at a geometric rate.
2 Food production increases at an arithmetic rate.

Therefore population growth will eventually exceed the capacity of agriculture to feed that growth. Preventive checks, such as delayed marriage, slow growth once the food supply ceiling is reached.

Positive checks, e.g. famine, disease and war, and increased death rate, keep the population in balance with the food supply.

In MEDCs Malthus's predictions were slow to be realised because:

→ they did not anticipate the Industrial Revolution
→ they did not foresee growth in agricultural output
→ transport and refrigeration helped move more food to markets
→ Malthus confused moral and religious issues with population issues
→ he could not predict medical advances such as birth control.

The figure below shows how Malthus's predictions are still not being achieved although the relationship between population and food supply is approaching a critical point.

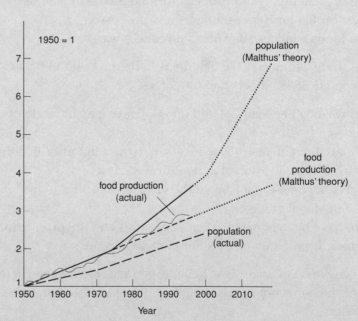

Boserup's theory, 1965

Society only progresses when it is under pressure. The greater the pressure of population and number of mouths to feed, the greater the stimulus to improve the food supply. Boserup was writing at the time when the Green Revolution began. Population stimulates economic growth because it forces people to discover new agricultural methods and crop strains with increased yields, to feed the extra numbers.

Action point

Write down some of the evidence of Malthusian checks to population growth that have occurred in LEDCs in recent years. Give actual examples of war, famine and disease.

"No economist, not even Marx or Adam Smith, has had his arguments so spectacularly misreported as Malthus. If he had been such a fool, Darwin would not have recognised him as his main inspiration."

Michael Lipton in a letter to the *Financial Times* 3 January 2000

"Necessity is the mother of invention."

Esther Boserup

Checkpoint 1

Write a sentence to expand on each of the points Malthus did not anticipate.

Malthus updated: The Club of Rome, 1972 ●●●

The **Limits to Growth Report** was prepared by the Club of Rome and is **neo-Malthusian**. Population growth is determined and limited by five factors: population, food production, natural resources, industrial production, and pollution. (Should we now, in the twenty-first century, add service industry output to this list?) The report concluded that if the five factors grew exponentially, the capacity of the earth to sustain population growth would be reached in 2070 when industrial (and service?) capacity might decline due to declining resources. To avoid this, policies of economic and ecological stability need to be introduced to bring about **global equilibrium**.

Equilibrium Ackerman's population: resource ratio ●●●

The theory looks at the relationship between the number of people and the quality of the natural resources and competence of the population. Types of population-resource regions:

→ The US and European type – advanced technological societies with low population growth and high availability of resources. Able to feed themselves and advance economically. Today these are recognised as MEDCs.
→ The Brazil type – technologically advancing but with high population growth and resources are becoming stretched, as industrialisation takes place. Some of these countries are NICs.
→ Arctic type – desert areas with low technology and limited food-producing resources. Today these are classified as LEDCs.

All countries strive to avoid the Malthusian trap of too many people and not enough resources to support the people. **Optimum population** or **optimum carrying capacity** is the ability of an area to support a population so that the natural resources are fully and sustainably utilised.

The jargon

Neo before a noun means 'a new or modern form of a theory'. *Neo-Malthusian* is therefore a new working of Malthusian theory.

Action point

Are there other countries that you could place in each of these three categories?

Checkpoint 2

What do *NIC* and *LEDC* stand for?

Exam questions answers: page 147

1 'Malthus is thought to have got things spectacularly wrong by claiming that populations would grow much faster than food supplies, causing starvation, misery and death, but failed to see the rise in food production as a result of science. Malthus sought to refute those who thought that the level of living of the poor could be raised by redistributing wealth which would result in more children and continued poverty. Population growth was soon to meet with checks on which Malthus placed limited hope.'
   Michael Lipton
   Why did Malthus place greater hope in schemes of improvement of land reform and crop production than in checks? (12 mins)

2 Describe some government policies to control population growth. (15 mins)

Examiner's secrets

Make sure that you have a range of examples of population control policies and migration policies selected from LEDCs and MEDCs.

Industrial location

Industrial location and the geography of industry are a core topic in geography. Industry was a core activity in the UK and it still is a major employer around the world. Therefore you need to know why industries locate where they are, how those locations are changing and what theories help you to understand the patterns of industrial location. Industry is the secondary sector.

Test yourself

Write down examples of jobs that are covered by the primary, secondary, tertiary and quaternary sectors of employment.

Changing importance of industry

The Clark-Fisher model illustrated below shows the changing importance of employment over time.

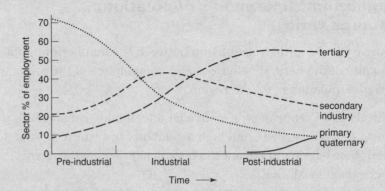

Action point

Place the following countries on the graph: UK, Poland, China, Germany, Malaysia, Kenya, Argentina. Write down why you decided where they should be positioned.

Theories of location

The original models of industrial location were developed by Weber in 1909 and refined by Losch in 1954. These are now just a starting point and are less relevant. **Weber** introduced some key terms:

→ **least-cost location:** the point where a product could be made most cheaply as a result of the interplay of raw material and transport costs from the resource areas
→ **weight-losing industry:** where product is less heavy than the raw materials
→ **labour costs:** the total costs of employing people to make a product
→ **agglomeration economies:** the savings made by locating close to specialist services and like firms needing similar skills.

Losch emphasised the role of demand from the market rather than raw materials. Brewing and baking are two traditionally market-based industries.

Smith's spatial margins model points out that industries do not locate at the optimum location because there are many locations from which profit can be obtained. Entrepreneurs have imperfect knowledge and select the most satisfactory locations – these are not the same as the most profitable. Smith moves towards a **behavioural approach** where those deciding locations go for what they know, and certain social, economic, political and business considerations that are imperfectly remembered.

Checkpoint 1

Give an example of each of the industrial location factors shown in **bold** type.

All these approaches are dated because:

→ industries today are **multi-product enterprises** with resources from many places and selling to many markets, e.g. car assembly
→ companies have more than one factory in more than one country and their HQ might be somewhere else – these are **multinational companies (MNCs)** or **transnational companies (TNCs)**
→ other companies are amalgams of a whole series of industries, e.g. Mitsubishi and BAT.

Industrial location factors ●●●

→ Raw materials – where they can be found, their proportion of the weight of the final product, material index, where and how they are moved, break of bulk points, e.g. at ports.
→ Energy led to coalfield locations in the Industrial Revolution but more recently electricity has enabled industry to be footloose. Some industries remained on coalfields because of industrial inertia.
→ Transport costs of moving, line costs, unloading, terminal costs.
→ Land can be a high cost if much is needed – cheap land sought.
→ Labour skills and quality, and labour relations.
→ Capital: the money invested in the production process, land, etc.
→ Government intervention and policies.
→ Behavioural reasons – the decision is made because of personal preferences (born there, golf courses).

Multinational companies have more complex locational decisions to make. To locate in the UK a company such as Nissan had to consider: 1. The need to be in the EU; 2. Other Japanese companies; 3. Attitude of local government; 4. Greenfield site; 5. English as a global language for business; and other factors from the list above, especially government assistance.

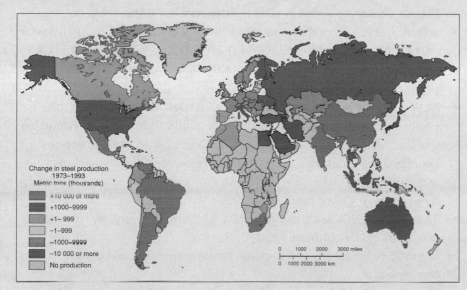

Change in steel production
1973–1993
Metric tons (thousands)

- +10 000 or more
- +1000–9999
- +1– 999
- −1–999
- −1000–9999
- −10 000 or more
- No production

0   1000   2000   3000 miles
0 1000 2000 3000 km

Exam question   answer: page 147

1   (a)   Describe where in the world the production of steel will have decreased and where it will have increased in the last 25 years. (12 mins)
    (b)   Explain the reasons for these changes. (12 mins)

The jargon

It is permissible to use MNC and TNC so long as you explain the abbreviation at the first mention.

Checkpoint 2

Name some *footloose* industries.

Examiner's secrets

You should have learned the location pattern of at least *two* industries. Iron and steel making is a good one for the basic principles. Car assembly is a good one to illustrate the more modern location factors and especially globalisation. The locational pattern and its changes should be known for the UK and the world.

Don't forget

Examples of government policies are:

→ Regional Development Policies
→ government assistance
→ tax concessions
→ Business Enterprise Scheme
→ investment policies
→ development Grants.

Government can also regulate industry through environmental laws, health and safety regulations, taxation, competition policy, labour laws.

Checkpoint 3

What kind of labour skills are needed for the electronics industry?

Examiner's secrets

In your answer consider the following factors:

→ Sources of ore
→ Transport costs
→ Labour costs
→ Levels of demand
→ State policies

Global shift

Action point

The following places are all leading global centres of technology. Select one and find out why it has become important: southern California, Boston MA, Seattle, M4 Corridor, Munich, Nice, Sophia Antipolis, Montpellier, Tokyo, Seoul, Singapore.

'Global shift' is a term introduced by Dicken to describe the locational shift of manufacturing and service activities across the world. It is a process very much controlled by multinational companies. Globalisation is present in many aspects of economic geography: companies have world brands, shops are managed from the USA, e.g. Gap. Goods are designed in one country, manufactured in another and assembled in another before being sold.

World manufacturing

Some 86% of manufacturing production is concentrated in 15 countries (USA 27%, Japan 21%, Germany 12% and UK 4%). The share in LEDCs is growing.

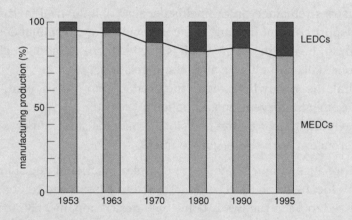

Why globalisation?

→ Labour costs are lower in labour-intensive industries, e.g. cotton spinning and weaving are mainly in LEDCs because labour is a major part of the cost and it is cheaper there. Making into clothing is also shifting to the low-cost countries, although some more expensive products are still made in MEDCs.

→ Skilled labour is found in countries with high skills but once the product is mass produced, production shifts to lower-cost countries – computers to Singapore and Taiwan and now to even lower-cost countries, e.g. Malaysia. Even more pronounced for microchip-making.

→ Need to have global markets – cars.

→ The role of TNCs such as Ford, IBM, Mitsubishi, Sony, which need new markets to continue expanding.

→ Demand of investors and pension funds for return on their investments.

→ The role of **tariff blocs**, which exclude imports but accept same product made inside their boundaries.

→ Organisation of companies to bring new products forward – product life-cycle is increasingly fast.

→ The role of technologies: IT and Internet.

Checkpoint 1

Select *one* industry from textiles, car assembly and electronics. Draw maps to show its changing global pattern of production.

Check the net

Company web sites might provide you with up-to-date information. A word of warning: remember that company sites are there to promote the company and will preach their own view of themselves and their activities – they are likely to be biased.

Product life-cycle model

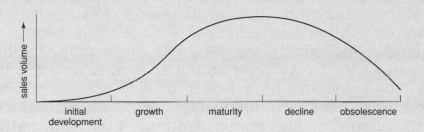

Initial development	Growth	Maturity	Decline	Obsolescence
Few buyers	More buyers	Peak	Declining	Fall-off
Low-tech	Mass production		Long production runs	
Capital small	High rate of investment		High – needs specialised equipment	
Know-how	More firms competing		Fewer firms as they leave	
R&D based	Management capital		Semi-skilled labour needed	
In EU/Japan	MEDC production	Shift to NICs	Shift to LEDCs	

Fordism and modern flexible production

Fordism is . . .	Flexible production is . . .
Complex technologies, production lines, high capital investment.	Flexible technologies, modules, easy to switch to new products.
Narrowly skilled workforce, with many semi-skilled.	Multi-skilled workers.
Suppliers distant. Just-in-case supplies.	Tiers of close suppliers. Just-in-time suppliers.
Very high volume.	Very high volume.
Standard designs with few variants.	Wide range of products. Lean – uses less of everything.

> "A transnational corporation is a firm which has the power to co-ordinate and control operations in more than one country, even if it does not own them."
>
> Dicken

The jargon

'Fordism is the development of assembly-line processes that permitted the production of large volumes of standardised products.' Dicken 1998

Just-in-time is the delivery of goods just when they are needed. This was developed by the Japanese as a means to reduce storage capacity.

Just-in-case is the delivery of goods and raw materials in such a way that they can be stored for use when they are needed. It is an extravagant use of space and has been replaced by just-in-time.

Checkpoint 2

Name some Fordist industries.

Checkpoint 3

Name a flexible production industry.

Exam question   answer: page 148

Explain how the product life cycle model can be used to explain the global distribution of consumer electronics. (15 mins)

Service industries

The term 'tertiary sector' is often used in the context of service industries. It comprises wholesaling, retailing, transport, the professions, leisure, public administration and personal service. It is based on goods. The quaternary sector is the sector that has been added by dividing up the service sector. It is the sector providing high levels of skill and education, such as research and development (R&D), financial management and administration. It is information based/knowledge based. Other classifications distinguish consumer services from producer services.

Action point

All these activities could be studied in a town. The theme could be where are they, why are they there and what impact do they have on the environment and people?

Checkpoint 1

What is *facilities management*?

Checkpoint 2

Why are there so many global centres in Europe?

The range of services

- Business services – facilities management
- Communication services – couriers
- Construction services – site preparation
- Education services – colleges
- Financial services – banking
- Health services – hospitals, vets
- Insurance services
- Personal services – day care
- Public administration – police, local government
- Recreational and cultural services – museums
- Trade services – hotels and restaurants
- Transport services – buses, travel agents

Globalisation of services

- Expansion of firms to gain a direct market in different countries
- IT enables transactions to be more rapid, e.g. bank transfers
- The expansion of TNCs
- Outsourcing, e.g. Swiss Air booking in Mumbai, India
- Financial services for the TNCs
- Global investments by companies
- New financial trading activities, e.g. securities houses
- Around-the-globe trading of shares
- Government deregulation

The main service centres around the globe are in a hierarchy of centres.

Type of centre	Examples
Dominant	New York, London, Tokyo
Major	Chicago, Paris, Singapore, Sao Paulo
Secondary	Toronto, Vienna, Hong Kong, Sydney, Johannesburg

These centres control most of the world's financial activity.

Services in cities

Retailing clusters in the CBD because there is a high threshold population. Customers come from a wide area to specialist shops – they have a large range. Shops cluster near one another or facing each other. Now retail clusters are often out-of-town where they can still attract a high threshold population. How do the big new shopping malls attract people? Shopping as a form of entertainment and leisure has revitalised markets. The larger the city, the more offices separate into distinct areas, such as banking, insurance, legal quarters, company headquarters, media companies and designers. All of these can locate in a specialist area – can you think of examples in a city that you know?

Finance centres of the top 100 companies

	No. of bank HQs 1996	No. of corporate HQs 1997	
Tokyo	13 (5)	18 (5)	International centre
New York	4 (2)	12 (1)	Supranational centre
Paris	10 (2)	11 (1)	International centre
London	5 (3)	3 (1)	Supranational centre
Frankfurt	6 (1)	2 (0)	International centre
Beijing	5 (1)	0	National centre
Osaka	4 (2)	7 (3)	National centre
Amsterdam	2 (1)	3 (0)	National centre
Zurich	2 (1)	2 (0)	International centre
Charlotte	2 (1)	0	Host centre
San Francisco	2 (1)	1	Host centre

What makes a financial centre competitive?

→ Agglomeration of demand, financial intermediaries and business organisations – size and production large
→ Large markets including multi-currency dealing
→ Culture of finance, expertise, contacts
→ Labour pool of talent in finance at a suitable cost
→ Discipline but not over-regulation
→ A central bank

Link

CBD, page 102.

The jargon

Threshold is the number of people required to make a service viable.

Checkpoint 3

Can you give examples of clusters? Why do some shops cluster and not others?

Project tip

An analysis of any of the functions listed in a city or large town could become the subject of a project. This is a good idea if you are studying economics or business studies.

The jargon

Supranational centres are places managing vast sums of foreign financial assets. Only two in world.
International centres are places where HQs of large international banks can be found. Only five in world.
Host centres are places that can attract foreign financial institutions.
Others are e.g. Singapore, Sydney, São Paulo, Rio de Janeiro, Chicago, Bombay.

Exam question

answer: page 148

Explain the pattern of dominant, major and secondary service centres shown in the table opposite. (10 mins)

Action point

Draw and retain for revision, a sketch map to show the location of service activities in a city that you know. Why are they located where they are?

Tourism

Tourism grew by 40% in the last decade of the twentieth century. It is now the largest employer in the world, with 1 in 16 workers dependent on it. It is growing by 7% annually. Geographers need to understand how it has grown, where people go and why, and, most importantly, the effects of tourism on the environment, economy and society.

Take note

In 1950, some 25.3 million people were tourists. By 1993 this figure had grown to 600 million.

Growth of tourism

This relates to transport and wealth. For example, in the UK:

Examiner's secrets

Remember that tourists do not only flow from Europe to the rest of the world. What is remote to Europeans may be the equivalent of the Mediterranean to Australians, e.g. Bali, Indonesia. Likewise the Caribbean is not a long distance for Americans.

→ Nineteenth-century – rail popularised coastal resorts. More affluent people took 'the Grand Tour' with Thomas Cook, an entrepreneur.
→ Inter-war rail and ferry travel to Europe – growth of early Mediterranean resorts catering primarily for own nationals.
→ Post-war – early air travel leading to European destinations; home packages – holiday camps.
→ Improved aircraft – package tourism; more holiday time.
→ Longer-distance tourism – larger aircraft, longer flight times, intercontinental travel; more time and money for weekend breaks; shipping technology leads to development of cruising.
→ Awareness of impact of development on environment – ecotourism.

Checkpoint 1

What is a *model*?

The Butler Model 1980

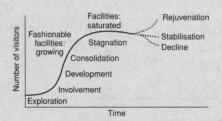

Types of tourism

Classified by:

Action point

For each of the types listed here, see if you can make notes on the various examples known to you, and the impact on the environment, economy and society.

→ type of environment – coast, mountain, heritage
→ type of interest – walking, skiing, fishing, sunbathing
→ length of stay – days, weeks, months
→ origin – home or foreign
→ wealth – package, exclusive, types of accommodation
→ season – winter sun or snow.

Alternative classifications are:

Checkpoint 2

Where is *Peter Island, British Virgin Islands*?

→ mass tourism to major resorts over 1–2 weeks – including skiing
→ weekend breaks, including city breaks
→ personally generated
→ business-related
→ ecotourism
→ cultural, including religious pilgrimages
→ sports tourism
→ touring

Action point

Remember that one place can have many types. Peter Island BVI has: day trippers from Puerto Rico; sailing and packages from USA and distant tourists from Europe.

- → short versus long distance
- → niche tourism – whale watching
- → educational
- → cruising – tropical, cultural, ecological and wilderness.

Impact of tourism

This could be looked at in terms of the environment, economy and society.

It could also include studies in LEDCs and MEDCs, or the impact on the countryside (coast) versus cities.

Environmental impact – trampling, path erosion and vehicle tracks; airport and resort construction; interference with longshore drift; rubbish left behind; carrying capacity; loss of land; endangered species or areas such as World Heritage Sites and SSSIs.

Economic impact – additional income; multiplier effect; employment multiplier; spending on travel, accommodation, food, local tours and car hire, souvenirs; impact on agriculture.

Social impact – employment; dangers of Westernisation of culture; illegal activities including sex tourism and drugs; migration out of rural areas.

Ecotourism and sustainable tourism ●●●

Ecotourism is also known as **green tourism** and **nature tourism**. It depends on nature and does not involve enhancing or improving nature. Why ecotourism? To find a new tourist market; to provide new, remote destinations; to place fewer demands on the environment; the effects of TV (David Bellamy in the UK); a search for the natural experience; the role of disciplines such as biology and geography.

Sustainable tourism is the product of the 1992 UN Conference on the Environment in Rio de Janeiro, *The Earth Summit*. **Agenda 21** was wider than tourism. It concerned itself with the threats of development to the environment. Sustainable tourism should and must:

- → conserve resources
- → reduce waste and over-consumption while supporting the local economy
- → promote cultural, economic, environmental and social diversity
- → develop a planning framework
- → involve consultation of people and communities
- → be marketed responsibly.

Check the net

Tourism Concern has a site at:
www.oneworld.org/tourconcern

The jargon

Ecotourism is ecologically and socially responsible nature-based tourism.

> "Sustainable development is development that meets the needs of the present without compromising the ability of future generations to meet their own needs."
>
> Brundtland Commission 1987

> "Sustainable tourism products are operated in harmony with the local environment, community and cultures, so that these become the permanent beneficiaries not the victims of tourist development."
>
> Agenda 21

Check the net

The World Wide Fund for Nature (WWF) is another major pressure group. Try their web site: www.wwf.org.uk

Exam questions   answers: page 148

1   Outline the costs and benefits of tourism to an MEDC and to an LEDC. (15 mins)

2   How has tourism changed over time? (15 mins)

3   What factors might hinder the development of tourism in an area? (10 mins)

Newly Industrialising Countries (NICs)

Link

See case study of NIC, pages 166–167.

The jargon

NICs is one term. Sometimes they are called *advanced developing countries*. Malaysia prefers *newly industrialising economies* because that reflects a broader set of achievements including service sector support for development.

Examiner's secrets

Rostow is merely a starting point for a discussion of the stages of development and does not include NICs. Make sure you know of some other models, for example the variant below which does adjust for NICs.

Action point

Can you write your own sequence of development for an NIC such as Taiwan or South Korea?

Action point

If Malaysia is not your chosen NIC, have a look at pages 166–7 where Singapore is used. Other NICs which could be used to prepare your own notes are: South Korea, Taiwan, Thailand, Mexico and (until it returned to China) Hong Kong. Some books include India and Brazil. Both of these countries have characteristics of both LEDCs and NICs, and most examiners will accept them as either, depending on the context.

Newly Industrialising Countries are relatively far along the path to industrialisation or a modern economy. They are the locations for much of modern industrial production, particularly in the fields of electronics and IT equipment.

Rostow's model of industrialisation

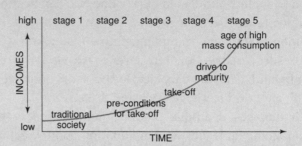

Malaysia's development to an NIC: Rostow revised

Phase 1 Early industrialisation The 1956 economy is geared to the export of rubber, palm oil and timber. Tin and oil are the main mineral exports. **Import substitution** – the production of goods needed by the population, e.g. clothing, to avoid expensive imports.

Phase 2a The new economic policy 1971–1990 Oil exports fuel growth. Government initiatives and central plans enable rapid growth of manufacturing + 26%. Creation of industrial estates; investment incentives, especially for Japan. In 1970, 60% of manufacturing was foreign owned. The process was government-led and not *laissez-faire* – a major departure from Rostow, market-led investment.

Phase 2b Push to heavy industrialisation and export-oriented electronics firms through state planning. Cars, steel, motorcycles, oil and gas. Tourism and education developing.

Phase 3 2020 vision Plan 1991–2000 and '2020 vision': government strategies to leapfrog Malaysia to developed world status. Emphasis on labour-intensive industries – shortages of skilled labour, high value added, high-technology, more sophisticated products. Routine assembly and less sophisticated production processes move to other South East Asian countries and China. In 1994, 360 Japanese companies in Malaysia.

The stages of European post-industrial experience depend on high-technology in all sectors, and will not pass through the labour-intensive service sector dominance as in the UK. There is an emphasis on state investment in huge projects such as the Multimedia Super Corridor, Labuan offshore banking, and Bakun Dam.

Where does Malaysia stand on Rostow's model? The country has progressed towards industrial status, and is described as an NIC by the World Bank. Malaysia prefers the title **newly industrialised economy (NIE)** whereas media describe it as an 'Asian Tiger', to convey an image of aggressive economic behaviour.

The Tiger Growth Model

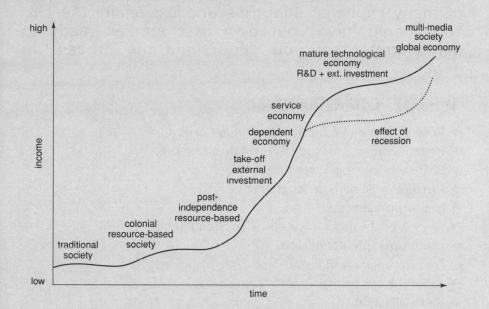

NICs as growth locations for manufacturing

Why growth?

→ Countries wished to substitute manufacturing for imports.
→ Able to attract foreign investment, especially from the regional leaders such as Japan in South East Asia.
→ Governmental creation of right conditions such as **EPZs**.
→ State enabled very large firms with a range of products – **Chaebol** in South Korea or **Zaiobatsu** in Japan.
→ Emphasis on consumer products that have high demand world-wide.
→ Strict labour laws, e.g. it is illegal to strike in Taiwan.
→ For the most successful, an ability to attract new investments in the tertiary sector and to invest in the region.
→ TNCs have tended to favour NICs, which has led to a clustering of similar activities, e.g. in Penang, Malaysia.

The social impacts of growth include a greater divergence in earnings, employment of women, use of ex-patriates, Westernisation of life, multiplier effect on, for example, purchasing patterns, growth in training related to NICs.

The jargon

EPZ is Economic Priority Zone. Many countries, especially in Asia, have a range of acronyms and abbreviations associated with all forms of development – *FIZ* (Free Investment Zone) and *FTZ* (Free Trade Zone) are two others used in Malaysia.

Action point

If you are using another NIC, draw up your own examples of each of these points. Also, have some detailed knowledge of the effects of development on people and the local environment.

Exam questions   answers: page 148

1   Explain why Rostow's model of development is less appropriate today. (8 mins)

2   To what extent have NICs begun to move to MEDC status? (12 mins)

Industrial decline

Industrial decline, which is sometimes called **deindus-trialisation**, is a key theme in the economic geography of many capitalist countries and especially the UK. You should have your own case studies ready to demonstrate that you understand the process in depth.

Why do regions decline?

→ Resources run out, for instance tin in Cornwall.
→ Better sources elsewhere, for instance iron ore.
→ Resources cheaper elsewhere.
→ Product cycle had run its course.
→ Newer products take over.
→ Growth of cheaper labour elsewhere (competition).
→ Downward spiral of neglect.
→ Less dependence of industry on coal for power.
→ Labour problems.
→ Rationalisation.
→ Image of area wrong for modern entrepreneurs.
→ Better sites in new areas.
→ Role of governments supporting growing industries, or competition for old industries.
→ Tax levels reduce profits.
→ Globalisation.

Cumulative causation in reverse – Myrdal adjusted is shown below.

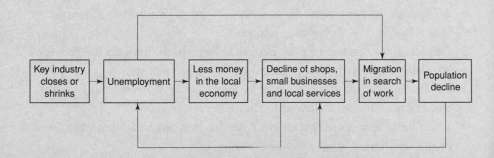

Declining regions

South Wales; The North East; Liverpool; The Ruhr, Germany; Sambre-Meuse coalfield, Belgium; Lorraine and Nord region, France; Rust Belt, USA.

Declining rural, remote regions is another category, often peripheral: Puglia, Italy; Central Massif, France; Central Wales.

The newly industrialising regions

These are replacing those above, e.g. the M4 Corridor, Heathrow to Bristol and into South Wales; Golden Triangle of Heathrow, Basingstoke and Guildford; Silicon Fen in the Cambridge area; Silicon Valley, California; Silicon Glen, central Scotland; Hessen, Germany; Provence, France. New growing industries prefer new areas with a dynamic image rather than declining regions that are more difficult to market.

Action point

Can you provide an example for each of these points?

Checkpoint 1

Can you draw the product life-cycle? Indicate on the diagram where declining regions should be placed.

Examiner's secrets

Try to have more than one region in decline to use. Everybody uses South Wales. Select another from this list.

Action point

A case study of a rising area might be possible, especially in terms of the broader effects on people and the environment. Another topic might be what to do with the derelict areas in declining regions.

Outcome: the spatial division of labour – low-paid jobs in one area and high-paid jobs in other areas.

Governments and industrial decline

Regional planning and various forms of regional policy:

→ Early attempts, such as Special Areas Act 1934, Barlow Commission 1940, Distribution of Industry Act 1944 and New Towns Act 1946
→ Special Development Areas and Development Areas
→ Development Agencies – Wales and Scotland
→ European funding, ERDF.

Project tip

The effects of government policies make excellent topics, especially if the project is at A-level rather than AS-level.

Urban-based policies

Post-1976. Frequent changes in titles and support. All attempt to solve the problems of decline by focusing on smaller areas:

→ Enterprise Zones
→ Urban Development Corporations (UDCs)
→ Urban Task Force Areas
→ Single Regeneration Grants
→ Business in the Community
→ City Action Teams and Task Forces
→ Inward investment authorities
→ Private enterprise new towns.

Action point

Can you describe the impact of any of these policies on an area near you?

European regional development policies

→ European Coal and Steel Community (ECSC).
→ European Regional Development Fund (ERDF) 1975.
→ Structural Funds 1994 (replaced ERDF): 'Objective 1' for areas with a low GDP; 'Objective 2' regions where economic change leads to industrial decline; 'Objective 5b' areas are rural areas in need of diversification.
→ Integrated Development Operations (IDO) – targeting a place.

Checkpoint 2

What does *GDP* stand for and what does it measure?

Exam questions   answers: page 149

1   Why have companies such as Sony come to South Wales? (10 mins)

2   Evaluate the relative success of one of the following: Urban Development Corporation, Enterprise Zone, Urban Task Force, Housing Action Trust, City Partnership Scheme. (15 mins)

Government involvement in economic development

The involvement of government in the economy is a key theme in the geography of economic development as well as of every economic activity. There are few aspects of people's lives where government does not have an impact!

Examiner's secrets

When geographers talk about government it can refer to almost any scale. At a global scale there are bodies such as GATT. At a continental scale the European Union has policies for every aspect of the economies of its members. At a national scale government is inevitably concerned with the well-being of the economy and the population. It will have policies for the economic sectors. At the local scale there are local authorities which also have controls over pollution and planning, for example.

Link

CAP, page 117.

Action point

Make your own notes on an example of each of the four types of intervention, naming a country and a case of how the government intervenes.
The case opposite is one for the UK but you should have at least one other from the UK and possibly one other country. Good alternatives are the north-east UK, Nord and Lorraine in France, the Saarland in Germany and Appalachia in the USA.

Checkpoint 1

What do you understand by peripheral region?

Supranational intervention

The General Agreement on Tariffs and Trade (GATT) was founded in 1947 and, subsequently, the World Trade Organisation (WTO) in 1995. By controlling trade it influences the distribution of activities. There are international pressures on governments to conform. The latest agreements were the Uruguay Round 1986–94 and the Seattle WTO 1999.

International policies – the European Union

→ Common Agricultural Policy (CAP) – late 1960s
→ Common Fisheries Policy controls resource exploitation and indirectly the jobs of both fishermen and those who process fish
→ European Monetary Union (EMU) and Euroland – creating common currency to enable goods to be exchanged more easily
→ Regional policies – ERDF

Types of governmental intervention

→ Market economies based on capitalism/competition/minimal interference
→ Market economies with regulation set by government – centralised planning may exist
→ Social-market economies where the state directs the economy but within economic and social aims including welfare
→ Socialist-planned economies where the state owns the means of production and controls wealth and income

Regional policy as governmental intervention

South Wales as a case of **deindustrialisation**:

→ Effect of **the depression** – designation of Special Areas
→ 1945–79 aid for industrial estates, office decentralisation, new roads, new towns and reclamation of derelict land
→ Continuing problems of the Valleys – loss of mining and strength of Cardiff as capital of Wales
→ Multinational and inward investments in south
→ Cardiff Bay UDC

Can you explain why growth has **polarised** in the coastal zone (see question 1 on the previous page)? Can you evaluate the success of the policies?

Peripheral regions often receive government aid. Brittany, or southern Italy (Il Mezzogiorno) and the Scottish Highlands are all good examples of peripheral regions.

Physical planning as government intervention ●●●

Structure Plans – post-1968 Town & Country Planning Act: broad-brush approach linking national needs, e.g. for housing, into a local context. Since 1990 in the UK the Secretary of State may intervene when policies are not consistent with national or regional policies. It is a framework for **District or Local plans**, applying the policies of the Structure Plan. Involves **Development Control**. People must be involved in the process (**public participation**). District Councils must have a Local Plan and within this Subject Plans for specific uses, e.g. marinas or retailing. The **Development Control Process** involves planners, politicians, developers and the public, often represented by **pressure groups**. These are voluntary bodies often starting out life as **interest groups**, which become pressure groups when their interests are threatened. Some are permanent, e.g. The Council for the Preservation of Rural England, whereas others form to oppose the building of a particular new road, for example.

Action point

Try to find out about a local planning issue. Pressure groups are often a good source of information for many projects. The local secretary of the group is the person to contact.

Other forms of planning control
These are all English examples but the same bodies are replicated in the other countries of the UK.

→ For Historic Buildings and Conservation Areas, the 1980 Local Government and Planning Act is the basis. **English Heritage**, a **QUANGO**, lists all buildings being conserved.
→ Nature Conservation – the 1981 Wildlife and Countryside Act created **SSSIs** (Sites of Special Scientific Interest), designated by the Nature Conservancy Council. **ESAs** (Environmentally Sensitive Areas), designated by the Ministry of Agriculture, Food and Fisheries (MAFF). **AONBs** (Areas of Outstanding Natural Beauty) are designated by the Countryside Commission.
→ National Parks – since 1949; a new park for the South Downs is currently proposed – do you know the others?

Jargon

QUANGO stands for Quasi Autonomous Non-Governmental Organisation.

Checkpoint 2

Name some of the National Parks in England and Wales.

These cases of government intervention extend beyond economic life but have an effect on the economic life of areas. You may wish to look at other cases in relation to geomorphology, such as **Heritage Coasts**.

Exam question answer: page 149

Study the following information.

> The old dock area around Cardiff Bay in South Wales is being redeveloped. The main elements of the plan are: a barrage across the bay to be turned into a freshwater lake; a marina in the old docks; a business park on the site of a former steelworks; shops and offices; new private housing; a hotel; a dual carriageway; parkland areas.

Who will gain and who will lose from the redevelopment? Give reasons for your opinions. (15 mins)

Development and disparity

Checkpoint 1

Draw the Rostow model of development.

Checkpoint 2

HPI bottom of the league (lowest first): Niger, Sierra Leone, Burkino Faso, Ethiopia, Mali, Cambodia, Mozambique, Guinea, Madagascar, Burundi.

Who is improving? Cuba, China, Zimbabwe.
Where is there little progress? Egypt, Guatemala, Morocco, Namibia, Pakistan.
Can you explain these patterns?

The contrasts between the more and less developed countries (MEDCs and LEDCs) are among the most stark that geographers study today. You need to know how development is measured, why it varies and what might be done to alter the patterns of development.

Where are the development disparities?

The simplest division between LEDCs in **The South**, and MEDCs in **The North**, is that put forward by Brandt (1980), which was originally part of a report making recommendations on aid, food production, trade, and energy consumption. Little international action followed.

Other indices have also defined the **development gap**:

→ **Human Development Index (HDI)** – development is a function of economic, social and demographic factors (UN data used on annual basis)
→ **Human Poverty Index (HPI)** focuses on the three elements of HDI, plus longevity, knowledge, and a decent standard of living
→ **Physical Quality of Life Index (PQLI)**
→ **Gross Domestic Product (GDP) per capita**, as shown in the figure below:

Country	GDP per head (US$)	Energy consumption per head (tonnes of coal equivalent)
Australia	14 000	7.24
Bangladesh	180	0.07
Brazil	2550	0.80
Canada	19 020	10.91
China	360	0.81
France	17 830	3.95
Ghana	380	0.11
India	350	0.31
Japan	23 730	4.03
Mali	260	0.03
UK	14 570	5.03
Zambia	390	0.20

→ **Social indicators** such as: students per teacher, literacy levels by gender, calorie intake, patients per doctor, persons per telephone, life expectancy, infant mortality rates, birth rates and natural increase of population.

Disparities are within countries as much as between them. The figure opposite shows the internal variation in a variety of characteristics of peoples homes in three countries (Quintiles mean fifths of the population). These figures are collected and published by the World Bank.

| | Quintiles | | | | | |
	Poorest	Second	Middle	Fourth	Richest	Total
Bangladesh						
If piped drinking water in residence	0.0%	0.0%	0.0%	0.1%	20.4%	4.1%
If uses septic tank or toilet	0.0%	0.0%	0.0%	3.4%	48.4%	10.4%
If has finished roof	0.0%	0.0%	0.0%	0.3%	31.0%	6.3%
Brazil						
If piped drinking water in residence	18.2%	61.3%	80.6%	87.6%	92.8%	68.1%
If uses a latrine with connection to sewer	1.4%	8.7%	13.4%	16.7%	11.2%	10.3%
If has cement or concrete roofing	0.3%	2.3%	13.2%	42.8%	94.8%	30.7%
Kenya						
If piped drinking water in residence	0.0%	0.8%	3.4%	27.6%	65.9%	19.5%
If has own flush toilet	0.0%	0.0%	0.0%	0.6%	32.4%	6.6%
If has roofing tiles	0.0%	0.0%	0.0%	0.1%	14.3%	2.9%

Explaining disparity

1 Friedmann's model of development is shown below:

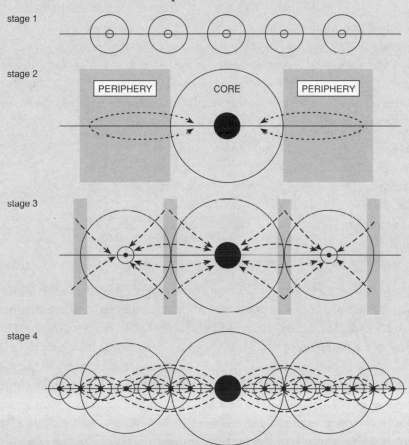

stage 1

stage 2

PERIPHERY   CORE   PERIPHERY

stage 3

stage 4

2 Myrdal's cumulative causation, which was used to look at decline on page 132

3 Frank's dependency theory

4 Marx's political model of development

Action point

List some of the inequalities that exist between regions of the UK. Some of these inequalities are more important to some people than others; just see if you can list which inequalities might be more important in the eyes of the Labour Party and the Conservatives. Which inequalities are highlighted by the Nationalist Parties in Wales and Scotland? NB There is a political dimension in terms of attitudes and values towards inequality.

Action point

What is *dependency theory*? Make your own notes on this theory.

Action point

Rework the Myrdal diagram on page 132 so that it explains growth rather than decline. Once you have done that you should have the original model rather than the adaptation.

Exam questions   answers: page 149

1 'Brandt's North and South is a gross oversimplification.' Discuss this statement. (15 mins)

2 Critically evaluate the various criteria used to measure disparity between countries and within countries. (15 mins)

3 Explain how the reaction of different governments to spatial disparities may be influenced by political ideology. (12 mins)

Trade and aid

The classic solutions to the issues of global development are those of trading so that poverty is eradicated and/or providing aid as the means by which poverty is eradicated. These are often painted as opposites in the press but, as a geographer, you will recognise that there are links between the two. A third force in development is the work of the **transnational corporations** (TNCs).

Checkpoint 1

Rostow devised his five-stage Model of Development in the 1950s. You drew the model as an explanation of development. Now explain the components of each stage. You should also be able to give an example of a country at each stage.

Trade ●●●

Trade is the exchange of goods and services.

LEDCs depend on exports of **primary products**, i.e. raw materials and cash crops. Prices of these have not risen as fast as the price of finished products that are bought in return, which leads to a debt problem.

Debt servicing – paying back the money and the interest – takes more than some countries can earn from exports. In 1994 it was three times all LEDC exports. Debt in sub-Saharan Africa was 14 times the amount that was spent on health and education in 1994.

Debt as % of exports 1995

Mozambique	1192%	Bolivia	410%
Ethiopia	458%	Indonesia	202%
Bangladesh	298%	Brazil	270%
Vietnam	396%	Mexico	170%
Nigeria	274%		

Source: *World Development Report 1997*, World Bank

Action point

Make your own notes to explain why one of these countries has such a debt crisis. The country or countries that you select should be those best suited to your syllabus.

Trade is dominated by OECD countries, where GNP levels are high, with large volumes of exports and imports, especially of manufactured products and services. MEDC/LEDC trade is unbalanced. The role in the past of **colonialism** is replaced today by the **neocolonialism** of TNCs.

LEDC/LEDC trade is still small. In some areas it is growing thanks to **trading blocs** such as **ASEAN** (Association of South East Asian Nations) – the Asian Tigers. Trade has enabled the oil-rich states of the Middle East to develop economically. However, social development here has not been as fast due to the cultural impact of Islam, e.g. the role of women. Why does the international trade approach to development have varied success?

Action point

Find out which countries are the members of ASEAN and LAFTA (Latin American Free Trade Association).

→ Uneven distribution of resources
→ Resource prices often stagnate or fall – role of TNCs
→ Markets for products stagnate so exports stagnate

Dependence on exports may result in a country's neglect of its own domestic production of essentials, for example cash crops instead of food. Imports have to substitute, reducing the impact of export earnings on development.

Self-sufficiency and development was attempted in India 1950–90. This meant that it was cut off from the world economy and it was slow, inefficient and bureaucratic.

Examiner's secrets

Brazil is a popular choice for many aspects of development. You might be able to score more by having a second case study that is African or Asian.

Aid ●●●

→ **Official development assistance** – one-to-one grants, loans and assistance
→ **Multilateral aid** – e.g. via UN, IMF
→ **Voluntary aid** by NGOs (non-governmental organisations), e.g. CAFOD, Oxfam

Much aid began for political ends, to stop the spread of communism in the 1950s. UN has aid goals but these only ever met by Norway, Denmark, Sweden and the Netherlands.

→ Budgetary aid – to help a country's annual budget
→ Non-project aid – covers debt relief, food aid and disaster relief
→ Programme aid – support for those with acute debt problems
→ Project aid – specific project support (e.g. dams) buys materials in exchange from the donor
→ Tied aid – may only buy goods and services from the donor
→ Untied aid – no limitations on what is done with aid

70% of aid returns to donors in interest and purchases. 10% of aid pays for debt relief and is rising. 6% pays for emergencies and has grown. In contrast, aid for education and health have declined as a percentage of total aid.

TNCs and development ●●●

Some 33% of all international investment comes from TNCs. Nearly all went to eight countries.

Cargill – an agricultural trading TNC based in the USA – supplies loans to farmers in Brazil to buy orange trees, purchases and process the fruit, exports in its own supertankers to the USA and sells to retailers. This has eliminated competitors. It puts cargo into financial futures markets to gamble on gaining more profit.

Trade in food results in low prices for producers but profits for processors and retailers. For example, cocoa prices are much lower than the price of chocolate when it is re-exported to the cocoa-producing country.

The top 12 TNCs by revenue in 1997 (there are other ways of measuring such as employment or turnover) were: 1. General Motors, USA; 2. Ford Motor Company, USA; 3. Mitsui, Japan; 4. Mitsubishi, Japan; 5. Itochu, Japan; 6. Royal Dutch Shell, UK/Netherlands; 7. Marubeni, Japan; 8. Exxon, USA; 9. Sumitomo, Japan; 10. Toyota, Japan; 11. Wal-Mart, USA; 12. General Electric Company (GEC), USA.

Aid is:

"The net flow of official development assistance (ODA) . . . the transfer of capital, usually in the form of loans or grants, from governments, international agencies and public institutions of the industrialised world to governments of the third world."

A–Z of World Development 1998

". . . a process where you collect money from poor people in rich countries and give it to the rich people in poor countries."

Schumacher 1974

"Transnational corporations possess and control means of production or services outside of the country in which they were established."

UN Centre for Transnational Corporations

Action point

Prepare your own case study of a TNC in e.g. car assembly or electronics. Banking TNCs could also form another section of your notes. Note where it was founded, how it has spread, how many are employed, and where and what is the impact on the new locations. The TNCs described here could be used.

Exam questions
answers: page 149

1 For any one TNC, describe the location of its main activities and explain how they have evolved. (15 mins)

2 What are the benefits and costs brought by TNCs in the countries in which they operate? (10 mins)

139

New global order

Examiner's secrets

Questions often use the word 'evaluate'. This is a high-level skill. You must be able to say what the good and bad points are about an idea and critically compare them with other ideas or solutions.

Much of the study of development looks at the impact of policies in the past on current patterns of development. However, one of the strengths of geography is that it does look forward and offers solutions to improve the world as the home of people. Prediction is a dangerous activity because it can be wrong when viewed 20 years later. Nevertheless, a good geographer should be prepared to look at the potential solutions to issues, and evaluate them.

Evolving solutions for global development

→ **Self-sufficient balanced growth** – China, India, Tanzania: very popular among socialist states in the 1950s and 1960s.

→ **GATT** 1948 – goods only, UNCTAD 1964 and WTO 1995 – goods and services: all promote free trade (see pages 138–39). Criticised for being too much influenced by TNCs in recent years.

→ **Trickle-down** – 1960s approach to poverty alleviation: wealth will gradually filter down to the poor (also used in 1980s UK with regard to regional disparities).

→ **Trade blocs** – LAFTA, EU, Comecon (1947–89 in Eastern Europe).

→ **NGOs** have played an increasing role since the 1980s. Not only involved in development but also in broader environmental and social issues. Have canvassed and protested at WTO meetings, e.g. Prague 2000.

→ **Bottom-up development** – projects should be aimed at the poor, especially in rural areas and on small-scale projects.

→ **Debt swap and debt relief** gained strength in 1980s – started by UNICEF. Debt for Nature swap by WWF and Ecuador, Madagascar and Zambia to preserve rainforests. Debt for Equity swaps in Brazil, Chile and Philippines – debt exchanged for shares in local companies. Debt for a footballer – PSV Eindhoven took over debt and bought Romario! Only 2% of debt is converted by these methods.

→ **Debt cancellation** 1990s – cancelled debt over a period of time on condition that saved funds are spent on education and health. Strings are attached. MEDC taxpayers foot the bill.

Test yourself

Evaluate the benefits to be gained from trickle-down and bottom-up approaches to development.

Tackling the social causes of inequality

→ **Demographic issues linked to poverty:**
  → high birth rates (Yemen, Ethiopia and Benin 49/000, UK 14, Germany, Italy, Spain 10)
  → infant mortality rates (Sierra Leone 163/000 live births, Uganda 120, Malawi 134, Finland 5, Japan 4, UK 6)

→ **Food and malnutrition issues:**
  → food import dependency (Ethiopia 15%, Sierra Leone 21%)
  → malnutrition – 800 million are malnourished (37% in Africa, 20% in Asia; in USA 30% of over-40s are obese!)

Examiner's secrets

Watch for outdated opinions. Roman Catholic European countries have some of the lowest birth rates and not the highest among MEDCs. Even in LEDCs the rate is not the highest.

→ Health issues
  → access to safe water (Ethiopia 25%, Sierra Leone 34%, Mozambique 34%, Myanmar 38%)
  → access to doctor or nurse, especially at birth
  → disease control – vaccines, new diseases, AIDS
→ **Education opportunity issues**
  → illiteracy – higher among females (adult illiteracy Sierra Leone F82% and M55%, India F62% and M35%)
  → enrolled in primary school – lower among females (Ethiopia F19% and M27%, Malawi F77% and M84%)
→ **Childhood exploitation issues**
  → sexual exploitation of children – sex tourism
  → child labour – 75 million aged 8–15 working to pay off family debts (16 million in India and 7 million in Brazil)
→ **Gender issues**
  → see health and education above
  → access to contraceptives – links to demographic issues above
  → deaths in pregnancy (Africa has 20% of world births and 40% of deaths in pregnancy and childbirth; Europe has 11% and 1%)
  → genital mutilation (100 million people) – leads to infections in up to 80% of cases
  → women not represented in higher-level jobs; tend to be in lower-paid and less skilled occupations even in MEDCs (in LEDCs can be tradition too, e.g. work in fields)

Agenda 21 ●●●

Product of Rio Earth Summit 1992 – six proposals for environmentally sound, sustainable development:

1 Support LEDCs in efforts to stop or modify projects known to harm the environment.
2 Give more resources to family planning, expanding job opportunities for women so that population growth is slowed.
3 Allocate development assistance to programmes targeting the alleviation of poverty, environmental health and meeting basic needs of food and shelter.
4 Invest in research on energy alternatives to reduce greenhouse gases and slow climatic change.
5 Invest in research and agricultural advice to reduce soil erosion and develop more environmentally sensitive agricultural policies.
6 Provide funds to encourage biodiversity and protect natural habitats.

Exam question   answer: page 150

Consider the World Commission on Environment and Development's definition of sustainable development on the right. Assess how far the societies of the world have moved towards this target. Explain why there has or has not been progress. (15 mins)

Link

Health issues, pages 152–7.

The jargon

Development gap is the term used by some examination boards to refer to the disparities between MEDCs and LEDCs and the potential of some countries to develop more rapidly than others.

Action point

Save the Children is concerned about child labour being used by some companies in South East Asia, e.g. Nike and Gap. Gather your own information on this issue.

Test yourself

To what extent has international trade been responsible for the development gap?
Suggest the effects of Agenda 21 on the development gap.

". . . a process of change in which the exploitation of resources, the direction of investments, the orientation of technological development and institutional change are all in harmony and enhance both current and future potential to meet human needs and aspirations. It is possible to obtain sustainable development."

The World Commission on Environment and Development

141

Answers
People, settlement and economy

Population: demography

Checkpoints

1 Choropleth, topological, dot distribution or proportional symbols, maps.
2 Highest densities in very small states and lower in the very large states in the less hospitable parts of the world. Low also where high living standards partly due to high resource availability. You need to be very specific for each country.

Exam questions

1 Start with the model and then state the demographic details of two countries at each stage, e.g. Denmark or Italy or UK at Stage 4. The details should include birth and death rates and the rate of natural increase with supporting figures. The data need not be absolutely accurate but should be close to the real figures.
2 (a) This graph is characteristic of a university town with a large number of both sexes in early adulthood.
   (b) is a community with a large influx of males – many resource frontier mining settlements are like this. Also characteristic of immigrant labour.
   (c) A retirement community very much in the USA model – no UK retirement areas have reached this state although wards in some south coast towns have a similar structure.
   (d) a larger number in early adulthood. This could be a place with married workers moving in.

Migration

Checkpoints

1 Jobs available here because close to Europe and because there are modern jobs, e.g. in finance and computer-related activities; higher-income prospects; decentralising from London.
2 Migration to sun belt and retirement migration. Job prospects in high-tech and defence industries. Move away from rust belt and declining industries. Poorer quality of life and environment in old industrial areas of north-east.

Exam questions

1 Your answer should include a classification, there are many in your textbooks. The second part of your answer must be some reasoning, with supporting examples, for the chosen classification. Justification does imply that the answer should say why you prefer a more sophisticated classification rather than a simpler one.
2 Government in this case refers mainly to national governments but it could also refer to local government. Growth management should relate to the policies discussed in the section on demographic issues, but see if you can think of the ways by which governments attempt to slow down population growth. The management of migration can range from forced migration and ethnic cleansing to policies that make migration difficult by not building houses, and instituting permits to live in an area, e.g. the Channel Islands. You must have more examples of policies, e.g. policies to welcome Commonwealth immigrants as in 1950s UK, or to the Gulf States. Local authority policies to restrict house-building in villages actually control migration.

Demographic issues

Checkpoints

1 The elderly are different because there are fewer of them, they are part of the extended family, they have a role in child care.
2 Note that it is *numbers* and not *proportion*: numbers because the post-war baby boom is now reaching retirement age. Welfare states and disease control enable longer and healthier lives.
3 Concentration into some areas – ghettos, schooling with English as a second language, pressures on local authorities to provide shelter, support issues, issues to do with development of cultural facilities such as temples and mosques and retail specialisms. Governments help by providing human and physical resources, by legally combating any racism that might occur, by attempting to disperse population, by producing documents in a variety of languages.

Exam questions

1 A good way to approach this question would be to start with a series of population pyramids for a country (UK?) from 1900 to 1991. From these you would be able to show that there have been changed demands for services. The post-1945 baby boom led to demands for schooling and later for house-building and e.g. New Towns. More recently the slowing down of population growth and smaller family size have led to an ageing society. The issues of caring for and supporting the elderly, ranging from migration to the provision of suitable facilities, should be discussed with the support of examples which could come from almost any area of the UK.
2 The processes are those of clustering of peoples in areas where they feel at home with 'like' people. Immigrant groups tend to start in key low-rental areas and then spread outwards, often displacing the former population. Displacement is the result of both 'white flight' and the new people being the only ones interested in the area. Some have moved to the outer suburbs as they have become more established. Low-rent areas are also close to industrial areas and the airport.

Urbanisation

Checkpoint

An urban centre which dominates all the other cities in a country with a population much larger than that of the second largest city, e.g. Paris, Dakar.

Exam questions

1 Time is one factor. Early megacities were in MEDCs but due to population growth slowing down and the processes of suburbanisation and counterurbanisation, these cities are stagnating and even declining. Very few are now in the largest category. The rapid urbanisation of cities in LEDCs has resulted in a large number of megacities in the tropics that have reached huge numbers in the past few decades. A second theme is to do with the nature of urbanisation in LEDCs – urbanisation without industrialisation. Demographic issues have also led to a high birth rate and natural increase among those in the cities. It is not all to do with migration; primacy of some cities such as Tehran and Jakarta, could be mentioned.

2 *Similarities*. The rapid growth at a period in history – nineteenth-century UK and twentieth-century India. Migration is a common theme at all stages. Youthful population with a high birth rate is another theme. Rapid spread in the physical extent of the city. Cheap housing. Poor sanitation.
Contrasts. Lack of industrial foundation in LEDCs. More dependent on services from outset. Scale is greater so housing shanties more common. Many have a reversed social structure – poor people on the outskirts. Inability of governments to cope also due to lack of funds and debt.

Rural settlement

Checkpoints

1 Due to decline of agricultural employment. To find work in urban areas. However, drift to towns is much less these days.

2 Bright lights lure them. Potential for jobs (often not reality, which is unemployment and underemployment). Pressure on land resource. Others have done so – family and friends.

Exam questions

1 Start by defining counterurbanisation and tell the examiner what your examples will be, i.e. the areas that you will be using. The effects on the *economic* geography will be a revival of trades doing up properties such as barn conversions, counterbalanced by loss of retailing because newcomers will use towns. Change to more antique shops and shops to cater for new and possibly temporary residents in second homes. House prices and rents forced up so that true rural population may not be able to live there. Less demand for public transport, which will make for difficulties for those without access to a car. *Social* effects will be a change in the class make-up and employment of the population. Some work provided in upkeep of properties may result in some people staying. Fewer young people and a slightly top-heavy population structure.

2 *In MEDCs*: has been going on since industrial revolution. Was a migration to work in towns. Depopulation now partly replaced by commuters. Land not abandoned.

In LEDCs: still very gender-biased (it was initially in MEDCs). Many retain interest in landholding. Land abandoned because had already been overcropped. Larger scale movement in LEDCs.

Internal structure of cities

Exam questions

1 This very much depends on your own case study. A sketch map should be drawn. The map should identify the CBD or central area together with any subdivisions, the existence of subsidiary centres, industrial zones and industrial estates, any zone in transition, the main housing areas by age and class characteristics. It is unlikely that Burgess will provide the explanation. It is good practice in this answer to demonstrate that UK towns and cities are similar to Mann's model or a mix of Hoyt and Burgess. You will also need to draw a sketch of the best-fit model.

2 (a) Criteria are such factors as income levels, education levels and provision, amount of open space, level of criminal behaviour and drug use, population density, age and upkeep of housing.

(b) *MEDCs*: quality rises with distance from the centre due to ability to live away from the workplace, suburbanisation, transport, affluence and ability to afford more space.
CIS (centrally planned economies) decline with distance because state invested in central housing which had greater political prestige, and kept down transport and infrastructure costs. However, the decline is slight because of the egalitarian nature of the socialist system.
LEDCs: best nearest centre and then rapid decline due to access and high status accorded to living in centre. Peripheral shanties for the immigrants have lower quality of life.

(c) You need to have a city up your sleeve. Best to take a UK city but not London, which is too big. You can show how cities conform to the basic model but there are exceptions such as those on the coast where high quality stretches along the coastal strip, enclaves of higher status close to the city, e.g. Georgian Bath or higher ground (Clifton, Bristol). Industry, motorways and ring roads may distort pattern.

Social geography of cities

Checkpoint

Social: Ageing population, increased poverty, homelessness and deprivation. More single parent families and ethnic minorities. Environmental: More derelict land, decline in quality of buildings.

Exam questions

1 (a) Slow growth for 80 years to 1950 followed by almost exponential growth between 1950 and 2000. You could add the figures.

(b) The percentage has increased and is over 50% in 2000. The number has shadowed total population and increased faster from 6 to 10 million in 10 years.

(c) Provides housing for immigrants. Housing costs are cheap. It is very close to many areas of work. Support system in favelas, a cushion for immigrants.

(d) You should have your own examples of the following: waste disposal leads to disease – running open sewers; erosion of slopes in tropical rains; slumping of slopes; air pollution from open cooking; water supply problems – pollution.

2 The question is about factors. Originally religious persecution. Immigration of cultural and ethnic groups. Need to be with kindred people, e.g. 500 000 Afro-Americans crammed into South side Chicago in 1930s = preservation of culture as well as being defensive. White flight, e.g. in Detroit, as fear of ethnic groups forces people away. The role of estate agents 'block busting', in promoting area to an ethnic group. Role of transport in spread. The role of certain cultural foci, e.g. mosques, supply of special foods, e.g. Hallal products. Spread by leapfrogging.

The central area and service sector

Checkpoint

You should be able to answer this for several examples. Consider why they have developed as they have and what problems these developments might cause.

Exam questions

1 (a) Mixture of ages in Manhattan from 1930s stepped blocks to glass-clad higher. Activities are offices for banks and service sector firms – insurance, company HQs.

(b) All buildings will be modern, post 1970. Usually glass clad not brick built.

2 In heart due to access for workers and for population using services. Need to cluster for external economies, such as recruitment and supplies of other services. In cities such as NY and Singapore because they are at key points in the time zones of the world. A global pattern of banking and control points for their respective regions.

Issues in cities

Checkpoint

There are more single-person households – elderly alone, divorced/separated, 'singletons' = single working persons living alone.

Exam question

(a) *Housing* by internal modernisation and extensions. *Streets* by removal of through traffic and partial pedestrianisation together with landscaping. Garages provided with back access routes. *General environment* improved by trees and ornamental beds, wires placed underground. Some redevelopment, e.g. slum clearance.

(b) The first part should give you some suggestions. Primary and secondary data are defined on page 172. Primary work would involve mapping of the improvements and measuring their impact through environmental impact analysis and sample surveys of residents to ascertain reactions to improvements. Secondary work could involve finding out what it was like before, from maps and photos. Census data could give some idea of the changed population. All census material is dated – currently 1991 census. Perhaps there were alternative plans for the area that were not used – another secondary source.

Leisure in cities

Checkpoints

1 Home-based leisure using Internet. Learning as leisure. More gyms and health clubs, beauty therapy activities growing. Shopping is seen by some as leisure as we spend surplus earnings.

2 More space, e.g. for parking to catch the casual visitor along city entry routes, easier access, cheaper land.

Exam questions

1 The question focuses on urban leisure, which can be tourism or recreational use or both. Your introduction to this essay should define terms and say what you will cover. It is better to cover in depth rather than give a superficial answer.

Factors are: affluence of all age groups enabling new recreational venues to be established, shorter working week, socio-demographic changes such as later marriage and childbirth, planning of new zones such as central clubbing and bar area and out-of-town cinema complexes. Better transport – cars and ability to pay for taxis (wealth again). Existence of universities drawing more young to inner city. Fashions can result in more leisure, e.g. biking resulting in cycleways. Fitness leading to more sports venues. All these should be supported by examples from a range of places, with a focus on your home town and one other city.

2 Much will depend on your choice of urban area. This question can be about a part of a city – e.g. a segment of London. Environmental impact could be on the greenbelt or on parks. It could be the impact of traffic, parking, noise, litter and even light pollution from stadia. Consider the impact of artificial environments. Is there an impact on wildlife?

3 Definitions would be useful for starters. They are both when the supply is of local, national and international importance, e.g. many of the sights of London or Paris. Sometimes they are planned as such, e.g. Covent Garden. They are used longer if they satisfy two demands. The planning problems are to do with access – tourist buses clogging up streets and spoiling the experience for more local visitors. Tourism requires hotels not too far away, which might destroy other local recreational amenities. Provision of cafés and restaurants, which have to cater

for two types of demand. Overcrowding of honeypots. Note that this question does not actually say 'in cities'. Therefore, you could look at other cases, such as the seaside resort, which may be a day trip for some and a holiday venue for others. This is a question where you can use your project work.

Non-renewable resources

Checkpoints

1 Soil = critical. Water = non-critical. Solar energy = non-critical. Forest = critical.
2 *Costs*: Transport of waste paper for people and the companies. Reprocessing costs and separation costs. If demand low, profits are low.
   Benefits: Less woodland/forest used, national imports from producer countries less. Self-esteem of saving waste.

Exam questions

1 This diagram shows how coal is formed.

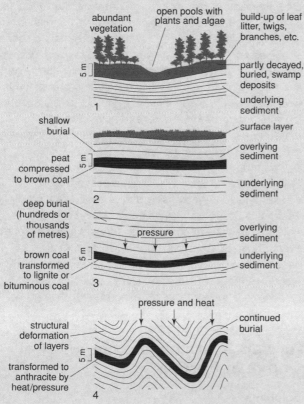

2 The first stage is probably a survey to ascertain the size of the resource, demand for the mineral, and the timescale over which it can be used at optimum levels. Bear in mind that profitability can rise as resource dwindles or falls in value and as new resources or substitutes are found. Most firms today will carry out an environmental impact analysis (EIA) of the site. It will also consider alternative sites, often on a global scale.
3 *Secondary fieldwork*: plotting the growing extent of the quarry from old maps. Some secondary work to gain knowledge about the company operating the quarry – whether they are part of a big group or merely a local

operator. The latter will be less intent on maximisation of gain over a short period.
Primary work could involve measuring noise from blasting and from increased traffic – best with official noise meters. Surveys of local population to ascertain objections to quarry or mine. Ask questions to find out if some people gain from the quarry, e.g. more people might be employed and others might be suppliers of equipment or even food.

Renewable and sustainable resources

Checkpoints

1 1. London Airport, but too late. 2. Ploughing downslope in many areas. 3. Where irrigation is taking place, e.g. Israel. More water control, such as drip feeding. 4. Mangrove destruction around Malaysia – stricter control e.g. on clearance for resorts. 5. Nepal for fuelwood – replanting schemes. 6. On savannas of Tanzania. Reduce number of animals but difficult because animals feed the population. 7. Acid rain on crops in northern Europe. Controls on power station emissions. 8. Water abstraction leads to wells drying up. 9. Loss of mangrove as habitat for many fish.
2 Principle 1: 'Human beings are at the centre of concerns for sustainable development. They are entitled to a healthy and productive life in harmony with nature.'
   Principle 8: 'To achieve sustainable development and a higher quality of life for all people, states should reduce and eliminate unsustainable patterns of production and consumption and promote appropriate demographic policies.'

Exam questions

1 Best to select a UK area although the Colorado river is another good case. In UK by increasing supply through boreholes and river abstraction, by re-use of water by more efficient waste management and recycling, e.g. Thames. By water storage in reservoirs, e.g. Kielder, and quarries, e.g. on Guernsey. Firms encouraged to develop own supplies that do not need purification, e.g. steelworks, and to recycle through cooling towers, e.g. Trent. Hosepipe bans, bricks in toilets.
2 Approximately: Oil 32%; Coal 26%; Gas 17%; Fuelwood 14%; HEP 6%; Nuclear 5%.
3 *For!* Save existing UK resources of coal, oil and gas. Save on expensive imports of energy. Lower running costs (but high capital costs at first). Environment is windy in west, seas can be rough.
   Against: Costs of installation. Solar unreliable due to climate. Wind and waves often not at the right time, e.g. cold snaps are generally calm. Wind power needs vast areas until generators more efficient. All small-scale at present.
4 (a) Population pressure, exploitation of forests to reduce national debt.

(b)

Advantages	Disadvantages
Help for forest dwellers. Will sustain forest through sustainable forestry. Debt will be reduced. Help for environmental movement. More exports with better markets.	Maybe not helping those who have been displaced. If a country does not accept the conditions, the people still lose out and suffer from forest destruction. It is still based on loans that might not be repaid and so debt increases.

Energy

Checkpoints

1 Wood, solar and wave are primary, diesel is secondary.
2 Increasing time to find wood because depletion has removed nearby sources. Other roles in society therefore suffer.
3 Not funded by the country but by loan from MEDCs. High interest rates. Costly external contractors because their know-how is costly. Aid is often tied to using external contractors.

Exam questions

1 It is the declining reserves of fossil fuels in the case of MEDCs and the decline of fuelwood in many LEDCs. It is a creeping crisis because as shortages arrive so prices rise and only the richest countries can afford energy. Crisis also caused by waste and inefficiency.
2 You can choose your countries.
   LEDC. For countries without fossil fuels it might mean increased consumption of renewable timber but preventing regrowth, e.g. Nepal. High cost of other fuels – debt. For those with some fossil fuels exporting to pay debts and not investing in the future, e.g. Nigeria. The theme is one of slowing economic growth and even stagnation and 'negative growth'.
   MEDC. Might have to invest in alternative supplies or expensive atomic energy, e.g. France, which has few home resources as coal and small oil and gas reserves run out.
3 You should deal with two if not three sources.
   Solar depends on good sunshine records throughout the year. Cost is also a deterrent, so there is support for the alternative view.
   Tidal needs an estuary with a good tidal range. Very few attempted, e.g. Rance, because of costs. There is an environmental issue to do with altering the physical geography of estuaries.
   Wave costly, environmental impact on area. Probably viable but other factors rule it out at present.
   Wind is unsightly although there is a magic to wind farms, e.g. Palm Springs. Again needs constant wind, so sites stand out. Costly to develop and still only supply relatively small amounts of power.

Geothermal still needs power stations. Possibly the most constant, e.g. Larderello in Italy.
All need to be backed up by other sources/examples.

Agriculture

Checkpoints

1 Need for cash crops to pay debts and aid development. Population pressures. Inefficient in eyes of capitalists. Natural threats such as drought, flood and fire to forests and grassland.
2 By subsidies, tax regimes, encouraging production, e.g. oilseed rape.

Exam questions

1 By clearing land and then clearing hedgerows and earth banks. By altering and simplifying ecosystems. By creating agribusinesses with large storage sheds, silos. By draining land. By breaking up iron pan. By eliminating pests.
2 *Drought* leads to crop failure, malnutrition, starvation and death. Even if last stages are not reached there is an effect on ability to work and to buy next season's seeds. *Diseases* attributed to malnutrition, e.g. beriberi, rickets and pellagra. Others, such as malaria, blackwater fever and sleeping sickness, all have debilitating effects on the working population and even more effect on subsistence farmers. *Pests* such as locusts remove crops and therefore can lead to hunger and starvation. Elephants can be a pest trampling crops.
3 Try to group reasons rather than be too emotive. *Physical reasons* to do with large areas of desert and semi-desert – increasing due to desertification. Poor soils. Droughts. Soil degradation and erosion. *Human reasons* – overpopulation, pressure on land and overcropping. Cash cropping. Human causes of desertification. Aid policies linked to those who have resources that MEDCs want. Debt burden increased forcing more to sell crops for export rather than grow food. Unable to afford modern crop strains.

Agro-ecosystems and agricultural change in Europe

Checkpoints

1 *Food chain*: Series of organisms with inter-related feeding habits, each organism serving as food for the next in the chain.
2 Fertilisers and pesticides remain in the system and/or can be moved from farm to farm by insects. Farmer needs time to re-invest.

Exam questions

1 Simplified ecosystems. Elimination of species of flora and fauna. Elimination of hedgerows. Pollution from sprays. Elimination of natural predators. Disease,

e.g. from sheep dips. BSE could be used but do not overplay this.

2 New crops – oilseed rape. Set-aside for golf courses, woodland. Monocropping more in 1970s. Effects on settlements such as rural depopulation. See page 115 for recent effects.

3 First define the types:

Organic farming	'Modern' farming
Inputs. Manure and nitrogen from legumes. Natural seeds from previous crop. Roaming animals.	*Inputs*. Genetically modified seeds. Artificial fertilisers, fungicides and pesticides. Animals housed.
Methods. Tolerance of weeds and pests. Greater use of labour, e.g. for weeding. Natural insemination.	*Methods*. Mechanical. Artificial insemination. High use of energy.
Outputs. Reduced. Products not so even. Due to demand, profits high and improving.	*Outputs*. Quality controlled for supermarkets. Profits high. Soil erosion and slurry issues.

Challenges for agriculture

Checkpoints

1 Food and Agricultural Organisation for the United Nations (its HQ is in Rome).
High yielding variety.

2 *Salinisation*: Process by which sodium, potassium and magnesium salts become concentrated in the soil.

3 Yields fall, land is unusable by machines, can cause disease in animals.

Exam questions

1 Depends on country but India and SE Asia provide some of the best case studies. Double and treble cropping leading to increased crop yields. Better able to feed population and reduce imports of food. Greater opportunity to export surplus, e.g. of rice and cash crops. Increased, expensive inputs of fertilisers and seeds.

2 Again you have a choice of country. A case study, e.g. Burkina Faso could be used. *Economic* life stopped due to lack of export crops and hides. Cotton did not grow. High cost of food aid – most paid for by MEDCs. *Social* should include demographic effects such as disease and death which have an impact on the economy. Orphans and elderly become a burden on society and may limit people's ability to work.

Can the world feed itself?

Checkpoints

1 Malthus lived before the Industrial Revolution. The agricultural revolution came later and, especially, the discovery of the New World. High-value goods are flown around the world, especially fruit out of season. Population control ethics vary between cultures.

2 Newly industrialising country. Less economically developed country.

Exam questions

1 The answer is hinted at in the quote. He placed faith in the schemes because they were within the grasp of people at that time. Land reform was taking place – enclosure and the end of open field agriculture. He did not forecast that food production would rise or that new lands would be discovered, e.g. North America, and that transport would enable food to be shipped around the world. Nor did he anticipate the growth in the use of contraception.

2 Policies could include the Chinese one-child policy that was state controlled. Singapore has a similar more voluntary policy. India's policies of taking advice to the community and using aid agencies to assist in the process is another. Legalisation of abortion can be seen as a policy. Policies to reduce pregnancies among the young through educational programmes, e.g. Netherlands, in contrast to unsuccessful UK policies to combat teenage pregnancies. Descriptions should record the outcomes in terms of changes in birth rates over time.

Industrial location

Checkpoints

1 Least-cost location = coastal steelworks.
Weight-losing industry = aluminium smelting.
Labour costs = high in MEDC, lower in LEDCs lower stages of development.
Agglomeration economy = IT-related industry in Thames Valley and Silicon Valley, California.

2 Toy making, electronics, furniture.

3 Mathematical and research skills, IT skills, programming skills, and small part assembly skills.

Exam question

1 (a) Growth in Asia, Latin America among LEDCs and NICs. In MEDCs Australia, Canada, S Europe and Finland are growing. Decline in the traditional industrial countries. Actual production has not grown – distribution has changed

(b) Production near the sources of ore, e.g. Australia and S Africa, Canada because lower transport costs. TNCs taking production to cheaper (transport and labour) locations. Rise of demand in developing countries such as China and India – production nearer the market. State policies to industrialise and steel is seen as the basic activity for industrial growth – the case in India, China and Brazil.

Global shift

Checkpoints

1 You need a case study of one of these industries.
2 Automobile assembly, electronic equipment assembly.
3 Modern auto assembly, e.g. the Golf floor plan used by 3 makes for 50 different models. Satellite assembly.

Exam question

Consumer electronics dominated by Asia. Historically three hearths of development: USA, Europe and Japan. The first two grew because of inventions, consumer demand and affluence, and production transferred to NICs and LEDCs where labour was cheap, with leading-edge production remaining. Japan followed a similar pattern but with a longer period of home-based production and development. Transferred out to NICs within the region. China has become the cheap labour location as costs have risen in NICs.

Semi-conductor plants are not found in the USA but in cheaper labour locations in Mexico and Caribbean. Heavy investment in Asian NICs, e.g. Intel in Malaysia – cheap labour. Product easy to move because small and light so no transport costs of note. Closer to main assembly plants. European investments partly to do with market (EU) access.

Service industries

Checkpoints

1 The maintenance and upkeep of a building and support for the activities inside it. It is a popular degree option replacing Quantity Surveying.
2 More developed economies, the result of a history of economic development in Europe, colonial past led to more controlling cities. Time zones enable offices to be open when both Asia and the Americas are at work as well.
3 Jewellers/shoe shops cluster for comparison.

Exam question

Dominant cities, London, New York and Tokyo have overlapping working days which was a factor during the early days of global banking. It helped establish dominance in the financial world. The major centres reflect dominance of MEDCs in providing services. In Europe it reflects the number of nation states. Only Sao Paulo in S Hemisphere. NB Singapore is a region centre for growing SE Asia. Secondary level reflects nation states but also growth of continental control centres e.g. Sydney, Johannesburg.

Tourism

Checkpoints

1 A model is a simplified representation of the real world to help understand a complex situation.
2 It is in the Caribbean British Virgin Islands.

Exam questions

1 You *must* name your country.
   MEDC. Costs include congestion, development of new hotels, a degree of stereotyping by tour operators. Benefits to balance of payments, multiplier effect, funds to maintain cultural heritage because tourists will be higher spenders. Improvements to infrastructure may assist local recreation and home-based tourism.
   LEDC. Costs in terms of leakages of funds, especially via TNC airlines and travel companies. Destruction of indigenous culture to fit a Western image of country. Unwanted activities, e.g. sex tourism. Diversion of agriculture to producing crops for visitors. Can rely too heavily on tourism at expense of other activities. Benefits include income for country, multiplier effect, diversifying employment, education for workforce, conservation of environment and heritage.

2 Changed because of transport, ability to move people and increasing affluence. UK Grand Tour as province of the rich in eighteenth and nineteenth centuries. Coastal resorts grew in nineteenth century with railways and paid holidays. Inter-war rail to Europe – ferries improved. Motor touring began for the more affluent. Post-war, holiday camps for mass tourism and growth of packages aided by air transport. Mass winter tourism followed. Long-haul tourism came with larger more efficient aircraft. Ecotourism as people wanted more than the package and could afford it. A realisation of what mass tourism was doing to the main destinations. Pattern different when viewed from other parts of the world. Barrier reef – mass tourism/Europe – long haul from Japan.

3 Depends on your area. Select an area where tourism is weak. Lack of access, lack of sites, hazards, difficult political environment, war, rise of alternative destinations, controls on access to prevent environmental degradation, overcrowding. Deal with both initial development and development from a particular point in time.

Newly industrialising countries (NICs)

Exam questions

1 All models are a product of the time they were written. Rostow assumed a standard, capitalist route to modernisation. Socialist states took a different route. NICs took different route as no need to pass through early industrial phases. New economic activities and role of TNCs.

2 You will need statistics like those on pages 166–7 for Singapore to be able to demonstrate that the country is similar to a MEDC. Some aspects might lag behind, especially social indicators. The answer is generally that NICs are closing on MEDCs in an economic sense but lagging in terms of social development. Because the question asks you to evaluate, you will need to have some contrasting NICs.

Industrial decline

Checkpoints

1 Refer to page 125.
2 GDP = Gross Domestic Product: Measure of total value of goods and services produced by a country, usually over a period of one year.

Exam questions

1 Work of a development agency (WDA) attracting firms, e.g. buildings, grants. The cluster effect of Japanese companies being together. The role of communications – M4. Reputation spreads among investing firms.
2 Have an example for the policy you select. Enterprise Zones have ended. Isle of Dogs and LDDC (Urban Development Corporation) have been overworked, so see if you can find out about another. Swansea valley is a good example. Manchester and Liverpool Task Forces. City Partnership, Newcastle.

Government involvement in economic development

Checkpoints

1 A region towards the edge of a country in terms of time and distance. It is remote from the heart of the country's economic and social life.
2 Lake District, Peak District, North Yorks Moors, Northumbria, Yorkshire Dales, Dartmoor, Exmoor, Pembrokeshire Coast, Brecon Beacons, Snowdonia, New Forest.

Exam question

Divide answer into the winners and losers.
Winners: (a) Professional people who will gain better job opportunities and luxury housing. (b) Watersports enthusiasts – new leisure facilties. (c) Local jobseekers – more possibilities of jobs in many fields. (d) Local businesses – new activity from workers/visitors will generate trade.
Losers: (a) Some local residents – may not be able to afford new housing and may feel excluded. (b) Environmentalists – ecosystem of the bay changed from a tidal one to freshwater, e.g. RSPB would protest about loss of habitat for wading birds. (c) Some local unemployed – no skills for new jobs.

Development and disparity

Checkpoints

1 See page 130.
2 Domination by the capitalist world and dependence of developing world. Profits taken by MEDCs/transnationals from resources and labour of developing world. Myrdal – capitalism would always produce dependence.

Exam questions

1 Crudely drawn line on the world map. NICs are put together with LEDCs – generalises across every LEDC from Mali to Singapore. It groups Ukraine with the USA and Germany. The reality is a gradation or a more complex set of types – former Communist Bloc, NICs, OPEC oil-rich countries. No account taken of size, e.g. small rich Kuwait as an LEDC. Line might be different if social development taken.
2 Discuss value of GNP, GDP, HDI, PQLI, labour force composition, exports by value, energy consumption, infant mortality, calorie intake, population per doctor, literacy, access to safe water as indicators with supporting evidence.
   Look at regional GDP, social indicators – telephones per 000, unemployment, population per doctor, education levels.
3 Right-wing governments place greater emphasis on market solving problems through trickle-down of benefits gained by developments in affluent regions. Left-wing places more emphasis on the state control of the means of production. Within that gross simplification have a range of examples that illustrate social market economies (UK and Germany), state capitalism (France), capitalism with a socialist state (China). Contrast governing parties' ideologies and their impact on spatial disparities, e.g. in health, unemployment.

Trade and aid

Checkpoint

1. Traditional subsistence society. 2. Pre-conditions for take off, trade, new inventions and technology introduced. 3. Take off, modern methods replace traditional. 4. Drive to maturity, a more complex urban society emerges. 5. Age of mass production and mass consumption.

Exam questions

1 Car assembly good case – Ford USA origins followed by expansion in Europe in inter-war period. After 1945 expansion within (Halewood, Saarlouis) and into new countries (Spain). Global sourcing of parts for just-in-time production in 1980s to meet competition from Japan. Bought companies (Jaguar), or acquired share (Mazda), spread global penetration.
2 *Benefits* – investment, jobs, development, exports, multiplier effect. *Costs* – leakage of profits to country of origin, dependence on overseas investment, control exercised from outside country – economically more powerful than government, changes to culture of country, immigration of skilled labour and managers.

New global order

Exam question

A question like this requires you to give some examples of ways that countries are moving towards sustainable development and obstacles to this progress and then summing up the overall state of affairs as you see it. The reasons for the progress and lack of it will probably be incorporated in what you write quite easily. You could include trade agreements, debt cancellation, population changes, 'green' policies, such as recycling, organic farming and renewable energy. You should also mention the increasing trend towards global co-operation as exemplified by the Rio and Tokyo summits as well as the less successful Hague summit of 2000.

Health and welfare

Health and welfare is a broad area within geography. It considers economic, social and environmental factors that influence a population's standard of living and quality of life. It has close links to population and economic geography and is important in the study of development. The topic identifies differences in levels of health and welfare by comparing and contrasting conditions in different places. It considers a range of causal factors and how these interlink. It outlines the effects of these on places and people and analyses some of the responses that may be made in order to improve conditions. An underlying theme is that, in their provision for health and welfare, countries need to adapt to constantly changing economic and social conditions.

Exam themes

→ Variations in levels of health and welfare at different scales, and indicators of this

→ Economic, social and environmental factors influencing differences in health and welfare

→ Types of disease and spatial variations in the incidence of different diseases

→ Transmission of disease

→ The effects of disease in different places

→ Responses to disease

→ The effect of demographic change on population structure

→ Identifying variations in life expectancy

→ Longevity and its causes

→ The impact of an ageing population and responses to it

Topic checklist

○ AS ● A2

	EDEXCEL		OCR		AQA		WJEC	
	A	B	A	B	A	B	A	B
Spatial variations in health and welfare	○●	●	○●	●	○●	●	●	●
Impacts of disease		●				●		
Spread of disease		●				●		
The elderly	○	●	○	○	○	○	○	○

Spatial variations in health and welfare

Jargon

'Per thousand' gives a good general comparison but important variations between age groups are not indicated.

Health is related to many factors including economic and social circumstances, genetics and behaviour. **Welfare** relates to the provision of services. The two are closely linked and vary greatly from place to place.

Health indicators and patterns

→ **Crude death rate** – number of deaths per thousand population
→ **Life expectancy** – average age at birth to which a person can be expected to live
→ **Infant mortality** – the number of babies who die before the age of one, per thousand live births

Variations in life expectancy by country are shown below.

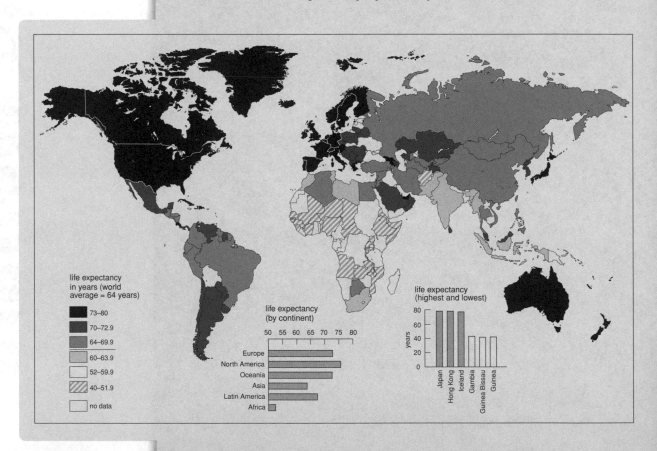

Checkpoint 1

Write a description of the patterns shown in the figure opposite.

Other spatial patterns

Inequalities will emerge at all scales such as:

→ between countries, e.g. within Europe
→ between regions, e.g. within UK
→ urban and rural areas (especially LEDCs)
→ within an urban area, e.g. within Greater London.

Factors that influence health and welfare

Some of the many factors influencing levels of health and welfare are shown below.

Population per doctor   Air quality

Access to clean water

Primary health care e.g. vaccinations   **Factors influencing health/welfare**   Social behaviour e.g. alcohol consumption, dietary habits

GNP

Personal wealth   Daily food supply   Medical research

Occupation   Goverment spending priorities

Allocation of funds

Variations in health and welfare depend greatly on capital available but the allocation of this money is also crucial.

Summary of the UK situation:

→ Government provides services through income from taxation.
→ Demand always exceeds supply, so there must be rationing. Within centrally determined limits, many spending decisions are devolved to local level.
→ Access to the same service, such as cancer treatment, will vary from place to place.
→ Faced with increasing demands on the NHS, pensions and social care, government increasingly encourages people to provide privately for their needs. The ability to make this provision is directly related to personal income, so wealth acts as a discriminator.

Scarce resources

Increasing wealth and innovation enable technological advances in healthcare. As standards of living rise, so do expectations. Thus resources will always be scarce. Poverty is likely to cause an individual to be more susceptible to illness and also to be less able to gain access to appropriate care, thus accentuating variations in health.

Exam questions   answers: page 160

1 Discuss the main factors that influence life expectancy. (15 mins)

2 'Access to healthcare is directly related to socio-economic position.' Discuss this statement. (15 mins)

Checkpoint 2

Choose one of these factors that will influence many of the others and briefly explain why this is so.

"To treat people with illness, disease or injury quickly, effectively and on the basis of need alone."

National Health Service policy statement

Examiner's secrets

Use supporting examples to illustrate your points. Comparisons between places are very valuable.

Impacts of disease

Disease has an impact on a society both socially and economically. Wealth plays a key role in a country's ability to prevent and treat disease. Disease has many *actual* effects but concern is also given to the *potential* effects of the spread of a fatal disease like AIDS.

Reducing the impact of disease – LEDCs

The main effect of disease is a reduction in the workforce and consequently in productivity. The costs of response measures are high and social stress is also a direct impact. A combination of preventative and medical measures is necessary in order to tackle disease. Many countries have benefited from the transfer of technology in the form of medicines, vaccinations and equipment, thus enabling them to reduce the incidence of disease despite their own limited income.

The variety of measures that can be taken are shown below.

Injection of money

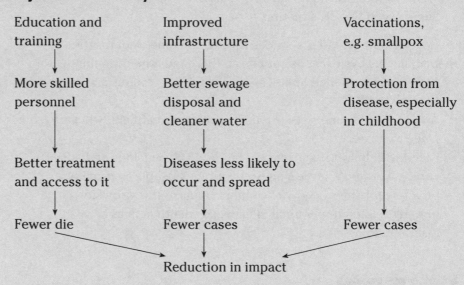

Education and training	Improved infrastructure	Vaccinations, e.g. smallpox
↓	↓	↓
More skilled personnel	Better sewage disposal and cleaner water	Protection from disease, especially in childhood
↓	↓	↓
Better treatment and access to it	Diseases less likely to occur and spread	
↓	↓	↓
Fewer die	Fewer cases	Fewer cases

Reduction in impact

Improved nutrition also increases resistance to disease, and specific measures like the spraying of mosquito-infested areas can also be effective. However, the control of disease requires ongoing attention and this involves a high cost that poorer countries cannot always meet. Thus there is still a wide incidence of entirely preventable diseases concentrated in LEDCs, and particularly Africa, that inhibits (with other factors) economic development.

The jargon

Disease can be tackled by trying to stop its occurrence; these measures are known as *preventative care*.
Treatment once the disease is contracted is known as *medical care*.

Checkpoint 1

What is meant by the term 'transfer of technology' and what might it involve?

Potential impacts of disease – AIDS in LEDCs

Some 80% of all AIDS cases occur in Africa where sexual intercourse without the use of a barrier method of contraception is common. Due to the long incubation period of HIV the disease is also likely to show a continued increase in the future. Over 20% of the populations of Zimbabwe and Zambia are thought to already be infected with HIV. If an epidemic occurs the workforce will be reduced and economic development held back. Less capital will be earned and so less money will be available for prevention and treatment. Thus the spread is likely to continue, further holding back the country's development.

Impacts of disease – MEDCs

Cardiovascular diseases including heart attacks and strokes are increasing in MEDCs. These are closely linked to increasingly sedentary occupations, rich diets and a lack of exercise. Deaths from cancers also contribute greatly to mortality rates, with other illnesses relating to use of alcohol and tobacco also of significance.

Many of these diseases can be reduced by a change in personal behaviour but the success of these initiatives is very dependent upon the response of the individual.

Such disease has a cost in the following ways:

→ health education to aid prevention
→ treatment – often complex and ongoing
→ counselling and support services
→ research and development; investment essential.

Despite long-running medical research, the hope of a cure for many diseases is still small. These include Parkinson's disease, multiple sclerosis, Alzheimers and many cancers. These are degenerative diseases in which a gradual and distressing decline in the quality of the sufferer's life is experienced.

More sensationally, concern is also raised about the emergence of new, and as yet untreatable, diseases like CJD and the re-emergence of drug-resistant strains of disease once thought to be controllable, such as TB.

Checkpoint 2

Using a flow diagram, show the potential effect of AIDS on a country.

Check the net

The UK NHS web site gives information about health and diet
http://www.nhsdirect.nhs.uk

Exam questions   answers: page 161

1   Explain how a lack of wealth can affect a country's ability to tackle disease. (15 mins)

2   Outline the main economic effects of disease. (10 mins)

Spread of disease

Disease can be broadly defined as any disorder of normal bodily function. It has physical, mental and social dimensions and these often overlap. Many diseases are preventable and treatable, whereas some are terminal and life can only be prolonged rather than saved. Identifying the origin, distribution and spread of diseases can help to prevent and treat them.

Origin and transmission of diseases

Many diseases are infectious – that is, they can be passed from one person to another. Most originate as a virus, e.g. HIV, or from bacteria, e.g. tuberculosis (TB), and to a lesser extent from parasites, e.g. malaria. Non-transmittable diseases include cancers and cardiovascular illnesses.

The figure below compares the causes of death in Africa and Europe (WHO member states).

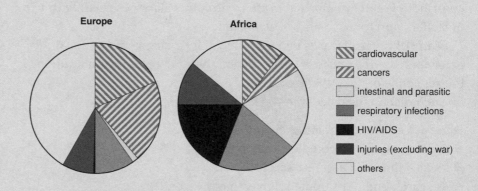

Check the net

The web site of the World Health Organisation (WHO) is full of statistics and information about health and welfare patterns and initiatives:
http://www.who.int

Types of transmission

→ Direct contact through blood or sexual fluids, e.g. HIV, syphilis, hepatitis
→ Contaminated or insufficient water, e.g. cholera, typhoid, leprosy
→ Insect vector, e.g. mosquito, tsetse-fly
→ Airborne in droplets, e.g. TB, whooping cough

Incubation period

This is the time taken for a disease to develop in a person before any visible signs emerge. Incubation for cholera can be only a few hours, malaria 10–40 days and HIV up to 10 years. Incubation time affects the rate of diagnosis and response and in some cases the risk of further infection.

The jargon

Diffusion refers to the spread of a disease over time and space.
Incidence of disease relates to numbers of cases.

Checkpoint 1

Why is it important to know about the incubation period of a disease?

The diagram below illustrates the transmission of cholera. Those living in poor and overcrowded conditions can easily be infected and an **epidemic** (rapid spread affecting significant numbers) can occur. Outbreaks regularly occur after severe floods and cyclones.

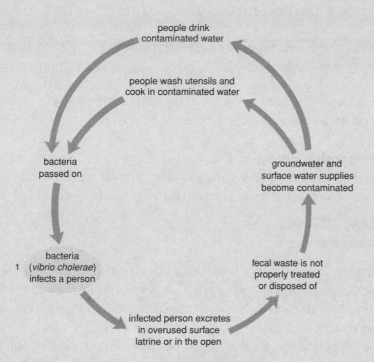

Don't forget!

Poor water supplies cause over three-quarters of all diseases in MEDCs. (Source: World Health Organisation)

Checkpoint 2

What is meant by the term *epidemic*? What conditions make an epidemic more likely to occur?

The effects of cholera are diarrhoea, vomiting, thirst, and muscle cramps. It can be treated with oral replacement fluids and antibiotics. If it is treated quickly, mortality is as low as 1%.

In contrast to cholera and other water- or air-borne diseases, the HIV virus (human immuno-deficiency virus) that can lead to acquired immune deficiency syndrome (AIDS) is spread much more directly. The predominant transfer is by direct sexual contact with an infected person and to a lesser extent by contaminated blood in transfusions or open wounds. It can also be passed from mother to baby.

Deaths from AIDS in 1998 totalled only 4.2% of total world disease mortalities but the potential for an epidemic in future is high because the long incubation period hides the extent of its spread at the current time. As yet there is no cure for AIDS, although life-prolonging drugs are available.

Exam questions answers: page 161

1 With reference to the figure on page 156, compare the incidence of diseases in Africa and Europe. (5 mins)

2 With reference to specific examples, outline the ways in which diseases can be spread. (8 mins)

The elderly

As economic and medical advances have occurred life expectancy has increased; this increase is termed **longevity**. An ageing population brings economic and social challenges and benefits to a country. The elderly are not a homogenous group; their needs are varied.

Nature of an ageing population

Ageing in a population has two main aspects:

→ an **absolute** (actual) increase in the *number* of elderly
→ a **relative** increase in the *proportion* of elderly

Demographic changes

An ageing population arises from longer survival rates (absolute and relative change) and fewer young (relative change), because of:

→ falling death rates
→ falling birth rates.

Increasing life expectancies are linked to many factors. Major influences are widespread prevention of many diseases, better prevention and treatment of disease, higher and more nutritious calorie intake, and better living conditions.

As education and contraception become more widely available, couples can exert greater choice over their family size. Societal change also plays a strong role. Fertility rates fell steadily in the twentieth century, and in more economically developed countries (MEDCs) birth rates are now close or equal to death rates. If they continue to fall, **replacement level** may not be achieved.

The combined effect of these demographic changes affects population structure. In addition, women live longer than men and so a gender difference also emerges.

The figure below illustrates the main aspects of an ageing population structure.

Watch out!

Make sure you understand the difference between *absolute* and *relative* change.

The jargon

Replacement level – on average at least one daughter must be produced by a woman during her reproductive years if a population is to maintain its numbers.

Checkpoint 1

Describe the main features of this figure.

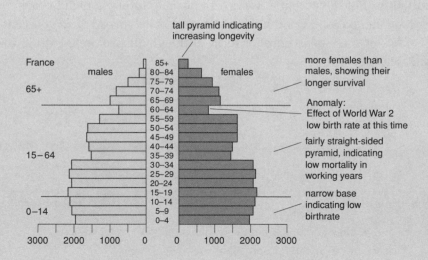

Providing for an ageing population ●●●

The **dependency ratio** indicates the relationship between the economically active or working population and the economically inactive. It is calculated as follows:

$$\frac{\text{Children } (0\text{--}14) + \text{elderly } (65+)}{\text{Population } 15\text{--}64} \times 100$$

Figures can range from 50 to over 100. A higher number indicates a higher dependency. In the UK the dependency ratio is gradually increasing as the elderly now make up 18% of the population and are projected to comprise 30% by 2030. Due to migration upon retirement even higher percentages are experienced in some areas, especially along the south coast.

Pensions

Some countries operate a 'pay as you go' system whereby those currently in work pay for the benefits of retirees. As the dependency ratio increases these benefit costs will rise and their payment will fall increasingly to a proportionately smaller 'working population'.

Strategies to cope with this include raising the retirement age, higher taxation, more **means testing**, and increased mandatory private provision. All these are controversial. In contrast many LEDCs have no pension provision and the elderly rely on their family for support.

Health and housing

In the UK the elderly account for a disproportionate amount of healthcare expenditure (over four times the per capita expenditure on those aged 15–64). There is often a shortage of state nursing homes and home services. More affluent pensioners are able to provide privately for their health and housing needs but those who can't become dependent upon an increasingly pressured public sector or other family members.

Valuing the elderly

The elderly are not simply a burden on the state; many undertake unpaid voluntary work for charities or act as carers for grandchildren. The 'grey market' is buoyant in products related to health, leisure and housing; this generates wealth.

Watch Out!

Dependency ratio figures can be misleading as retirement ages and working ages vary greatly in different countries.

Checkpoint 2

Why do some areas have a higher dependency ratio than others?

The jargon

Means testing is a method of allocating resources according to predetermined criteria. It is designed to target those most in need.

Exam questions answers: page 162

1   What is an ageing population? Outline its causes. (5 mins)

2   What economic and social issues arise from an ageing population in MEDCs? (30 mins)

Answers
Health and welfare

Spatial variations in health and welfare

Checkpoints

1 The majority of Africa falls into the two lowest categories with life expectancy below 60, e.g. Kenya, Nigeria. Northern Africa has higher levels of up to 69, e.g. Egypt, and in the south Botswana is similar. Asia has fewer countries with the lowest expectancies with the Middle East mainly showing ages of 64 and higher with South East Asia being somewhat lower. With one exception (Bolivia), North and South America have life expectancies of above 64 but the USA and Canada are the highest at 73+. This high rate is also seen in most of Europe, e.g. France and Germany with life expectancies decreasing eastwards. Australia, New Zealand and Japan are comparable with Europe. Overall, clear variations by continent are apparent, but change with latitude is less obvious.

2 The best choice is either GNP or personal wealth. The money available either to a country or to an individual is a major factor influencing the ability to provide and obtain key services. For example, if a country is poor it may not be able to ensure an adequate food supply and is less likely to be able to provide medical services, so the population is less likely to be healthy.

Exam questions

1 Begin by defining the term *life expectancy*. Give some idea of how it varies by giving some information from the map. State that there are many influencing factors affecting life expectancy that can be broadly divided into economic, social and environmental factors – this gives a clear structure for your answer. Choose a variety of factors to describe and for each explain clearly *how* they influence life expectancy.

You can then begin to make links *between* the factors, e.g. if children are undernourished and not receiving effective primary healthcare like vaccinations, they are more likely to fall ill as they are less resistant to disease. If the patient–doctor ratio is high and the region lacks medicines the sick child is less likely to receive adequate medical care and thus more likely to die.

It is also worth commenting on the relative importance of factors affecting life expectancy, e.g. a person's occupation is less influential than their access to clean water.

In conclusion, try to comment on the main causal factors, e.g. a lot of these factors link to availability of wealth.

2 Begin by showing what you mean by access to healthcare and socio-economic position. Add a comment reflecting the view that whilst a person's income and occupation can influence their access to healthcare, there are other influences too, such as where a person lives, and spending decisions made by health providers.

Show the country/scale at which you intend to answer the question – this will help to keep it manageable. The following relates to the UK. Give information about how in theory a person's income should not affect their access to healthcare, because one of the main aims of the NHS is to provide free access based on need alone. People pay for this service through taxation and are entitled to use it whenever the need arises. However, due to increasing demands for healthcare, it is clear that the NHS cannot provide adequate healthcare for everyone, i.e. demand exceeds supply. Therefore decisions taken about which services to offer will affect the aim of free and fair access, and some people will miss out.

Go on to show that people with different illnesses may not all be treated, and that decision-making will vary from place to place, so in this respect socio-economic position is less important.

Your final section could discuss how increasing pressure on the NHS has encouraged the more affluent to 'go private', thus accessing the care they need and receiving it more quickly. Income *is* a key factor here so it can be said that wealth affects access.

Impacts of disease

Checkpoints

1 *Transfer of technology* is a term used to describe the movement of innovations (and new ideas) from region to region or from country to country. It is most commonly associated with the transfer from MEDCs to LEDCs of goods such as machinery and tools, medicines and vaccinations. This transfer may be through government initiatives, charities, or increasingly through transnational corporations. It enables the recipient to benefit from an innovation to which they did not contribute.

2 HIV is spread among a population

After 5–10 years many AIDS cases emerge

Many workers become ill and die

The economy suffers

Lack of money to treat and prevent spread of the disease

The disease continues to spread

Increased mortality rate

The economy continues to decline

Development held back

Exam questions

1 Begin by stating that tackling disease needs many strategies. It is useful to distinguish between preventative measures (such as vaccinations) and treatment measures (such as antibiotics and surgery). Make the key point that all measures have a cost attached. Outline some of the strategies that can be undertaken but point out that whilst the solutions are often relatively straightforward (e.g. pit latrines), so many of them are needed that it is difficult to provide the required changes quickly. This means that although some areas will be improved, diseases will continue to spread in other areas. (You can usefully bring in knowledge from other topics relating to development and housing conditions here.)

The fact that many schemes need to be repeated, e.g. spraying of malaria-infested areas, may limit their effectiveness on cost grounds.

The potential spread of HIV in LEDCs could be linked to poverty, as condoms are not widely available. In conclusion, show that by tackling problems from a variety of angles the overall impact of disease can be reduced. However, unless capital (money) is made available from other sources, many LEDCs still find it difficult to implement comprehensive change as they are underdeveloped and their populations are still growing rapidly.

2 Begin by stating that disease has a cost to a society both in terms of responding to the diseases and in the resultant underproductive workforce; this gives a useful structure.

Outline the ways in which money is spent on strategies to reduce disease such as vaccinations, public health schemes, antibiotics, etc; remember to talk about preventing *and* treating disease. Point out that both rich and poor countries have to pay, because different diseases emerge as countries develop.

The second section should focus on how disease can hold back the development of a country not just by the costs outlined above but due to the reduction in workforce productivity as a result of illnesses. The possible effects of an AIDS epidemic could usefully be given here.

Spread of disease

Checkpoints

1 If the visible effects of a disease appear very quickly (i.e. short incubation time) it is likely that a more rapid response can be made and action taken to curtail the spread, e.g. by isolation. In cases like HIV the incubation period is longer and so the disease has much longer to take hold in a population and therefore can affect more people.

2 An epidemic is a rapid outbreak of disease that affects many people in an area at the same time. It is more likely to occur in densely populated areas with substandard living conditions allowing the disease to reach many people in a short time. After a natural hazard, e.g. cyclone or earthquake, a breakdown in essential supplies encourages diseases to spread, especially water-borne ones such as cholera.

Exam questions

1 This is a skills question and should not take more than 5 minutes.

Take each of the types of disease in turn and say how the Proportions are different. Do not simply give the figure; try to manipulate it in some way, e.g. the percentage dying from cancers in Europe is five times that in Africa.

2 This is a straightforward question, but make sure you only write about transmittable diseases rather than ones acquired in other ways.

Describe in turn the different means of transmission, i.e. air-borne, water-borne, etc. Make sure you give examples to support each one. Using a flow diagram like the one for cholera can show a sequence clearly.

Use terminology like *incubation* and *vectors* to show your knowledge.

It is worthwhile including a concluding point about how some diseases are spread due to poor living conditions (e.g. TB, cholera) as compared with those spread by certain social behaviour (e.g. HIV).

The elderly

Checkpoints

1 The graph illustrates an ageing population – the pyramid is tall and 'bell'-shaped. The narrow base indicates a low birth rate. The working age groups are evenly balanced and there is little gender variation. Above 64, however, there are proportionately more women.

2 The dependency ratio varies from place to place in relation to different demographic trends. A high birth rate and low life expectancy (common in LEDCs) tends to give a high dependency ratio as large family size and limited survival of adults into middle age creates many dependants to be supported by each person who is 'economically active'. As a country develops its dependency ratio will fall as the birth rate declines and life expectancy increases, giving proportionately fewer dependants per worker. However, as a country moves into an advanced stage of development, e.g. Sweden, the ratio may rise again as the combination of a sustained low birth rate and longer life expectancy will eventually result in fewer 'economically active' supporting proportionately more elderly. At a smaller scale the dependency ratio may be relatively high due to concentrations of elderly people in popular retirement areas, or relatively low in core economic areas where many economically active people live.

Exam questions

1 Define clearly the meaning of *ageing population*, by referring to both absolute and relative change, i.e. an ageing population occurs when there is an increase in the number of people over the age of 65 and when this group represents an increasing proportion of the total population.

　　Go on to describe the causes. The factors can be easily divided into those affecting longevity and those affecting the birth rate – this will provide an order for your ideas.

　　Do not go into too much detail as the command word is 'outline', not 'discuss'.

　　Conclude with the statement that it is the *combination* of the falling birth rate and increasing life expectancy that produces the ageing population.

Examiner's secrets

Give some examples of countries where the population is ageing. An annotated diagram of a population pyramid could save you time if you are rushed.

2 This is a wide-ranging question, so make clear in the introduction what you are going to write about.

　　Key economic issues are the provision of pensions and healthcare. Social issues relate to unequal access, burden of care, and society's perception of the elderly.

　　Discuss the increasing cost of providing pensions and healthcare as the proportion of the elderly increases.

　　It is important to show that their healthcare takes up a disproportionate amount of money. Develop your answer into a discussion of how to fund the increased demands. The social issues are the resultant variations in quality of life and the burden of care placed on families as their relatives age. This can be linked to the elderly generally being perceived as a burden – an example of negative stereotyping.

Examiner's secrets

This question is about issues, i.e. something over which people disagree. Keep your answer focused on conflicts over how best to provide, rather than just listing what needs the elderly have.

Selecting countries

The use of examples at a variety of scales is an important tool for geographers. One of the most important scales for human geographers is the national scale where examples are drawn from countries. Countries are almost always grouped into More Economically Developed Countries (MEDCs), Less Economically Developed Countries (LEDCs), Newly Industrialising Countries (NICs) and sometimes, Socialist and former Socialist states. The first three feature most strongly, and the next three spreads present some selective points that can be used in conjunction with the thematic spreads of the book.

Exam themes

The themes will depend very much on the topic being studied or answered. The key themes are the economic and social changes that are occurring in the country. The themes developed here could be developed in other exemplar countries.

Topic checklist

○ AS ● A2	EDEXCEL		OCR		AQA		WJEC		NI
	A	B	A	B	A	B	A	B	
LEDCs	○●	○●	○●	○●	○●	○●	○●	○●	○●
Singapore as an NIC	○●	○●	○●	○●	○●	○●	○●	○●	○●
MEDCs	○●	○●	○●	○●	○●	○●	○●	○●	○●

LEDCs

Action point

Draw basic sketch maps of your chosen LEDC showing population density and the main economic regions. You should be able to illustrate many of the themes of your course with examples from an LEDC.

For all of your answers involving different levels of development it will pay to have at least one country that you know well and a second to support your points in a more general fashion. Much depends on the syllabus that you are studying. We have used the term LEDC throughout this book; it stands for Less Economically Developed Country.

Kenya

Kenya is a former British colony which became an independent member of the Commonwealth in 1963.

Population: 26 million in 1994, which is expected to rise to 31 million by 2001. Growth rate of 2.8% 1992–2000. Urban 26%. 1.7 million people in Nairobi. An average of 5.1 children per woman. 1 doctor per 20 000; 90/000 deaths among under 5s. 91% of age group attend primary school, 23% attend secondary school. Per capita GNP in 1994 was $250 and stagnant. Debt is larger than GNP at $280 per capita. Development aid is 16% of GNP.

Regions

→ Coastal plain with tropical forest – location of the tourist resorts, and Mombasa. Water pollution is a problem, as are water supplies.

→ Inland savanna region with low rainfall including Tsavo National Park. Desertification a problem, especially in the north.

→ Central mountainous area with volcanic soils and a cooler climate provides good agricultural land – most of the economic activities are here. Soil exhaustion, due to population pressure, is a problem.

→ The west leading to the shores of Lake Victoria where 81% work in agriculture – cash crops (coffee, tea, sugar), vegetables, pyrethrum and skins. Some 7% work in industry – exports cement and chemicals.

→ Some 12% work in services – government, education, health and tourism.

→ The region has begun to benefit from ecotourism in Tsavo, Masai Mara, Amboseli National Reserves and Lake Nakuru National Park.

→ Threats to the region include soil erosion and degradation, a lack of resources, and AIDS, especially in the capital and spread along trade routes and rail links. Kenya has the third highest AIDS rate in the world.

Action point

If you are studying tourism, Kenya is a good example of several types of tourism ranging from package beach holidays to ecotourism. Have your own notes on the costs and benefits to the country of tourism.

Checkpoint 1

What is *ecotourism*?

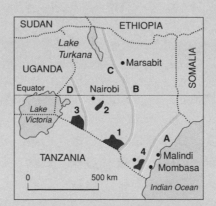

1 Amboseli Masai National Park
2 Lake Nakuru National Park
3 Masai Mara National Reserve
4 Tsavo National Park

A Coastal plain
B Savanna region
C Central mountainous region
D Western area

India

India is a former British colony that became independent in 1947 and was partitioned from Pakistan (including modern Bangladesh).
Population: 913 million in 1994, which is expected to rise to 1016 million in 2000. Urban 26%, Mumbai 12.6 million, Calcutta 11.1 million, Chennai 3.9 million. An average of 3.7 children per woman. 1 doctor per 2400. Under 5 mortality 119 per 1000. Maternal mortality of 437/100 000 live births. 430 000 cumulative AIDS cases to 1999, of whom 330 000 have died. In 1997, 140 000 died of AIDS. There are 120 000 orphans as a result of AIDS.

Illiteracy 48%. Some 91% of age group attend primary school, while 38% of girls and 59% of boys attend secondary school. Per capita GNP is $320 and growing 2.9% per annum. Debt service is 27% of exports, or 1% of GNP.

→ Monsoonal climate – hurricanes and typhoons are a threat in the Bay of Bengal – Orissa 1999 and 2001.

→ Living on the land – 62% work in agriculture, and village life is still the basis. Very small landholdings – under 2 ha, many sharecroppers own no land and pay rent from crops, which results in low investment and poor yields. The Green Revolution has helped to improve yields.

→ Industry – 11% work in industry – iron and steel, Damodar Valley; chemicals – Bhopal disaster in 1984; cars, food processing, textiles. Today Bangalore is home to major software producers.

→ Services – 27% work in services – government, etc.; tourism is a small employer.

Threats to the region include:

→ urbanisation – bustees, high densities, unemployment, prostitution
→ environmental – failure of monsoon, typhoons, pollution, water supplies
→ demographic – even though the rate of population increase is slower, the sheer numbers dictate rapid though slowing growth backed up by improving health and longer lives.

Exam questions

answers: page 170

1   With reference to one LEDC, outline the social, economic and environmental threats to its development. (12 mins)

2   Study the model of the interlocked cycles of poverty shown below. Explain how this model relates to an LEDC that you have studied (15 mins)

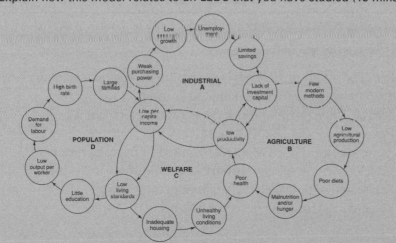

Checkpoint 2

What is a *bustee*?

Singapore as an NIC

Action point

It is worth having a bank of data to use in an introduction to an NIC. This list is an example:
Population: 2.9 million
Urban population: 100%
Children per woman: 1.8
>5 yrs old mortality: 6/000
Work in agriculture: 0%
Work in industry: 36%
Work in service sector: 64%
15% Malay
76% Chinese
6% Indian

Link

Malaysia, page 130

Don't forget

→ 1819 founded by Sir Stamford Raffles
→ 1867 British colony
→ 1963 independent as a part of Malaysia
→ 1965 independent state

Checkpoint 1

Make sure you know the difference between *GDP* and *GNP* and that you can correctly define *infant mortality*.

The jargon

Entrepôt is defined by J Small and M Witherick as 'A centre to which goods in transit are brought for temporary storage and re-export. A port where goods are received and deposited, free of duty, for export to another country. The synonym *free port* is being increasingly used.' The quote was taken from *A Modern Dictionary of Geography* by J Small and M Witherick, 1992. It is useful to have such a dictionary to help you learn the real meaning of the terms and concepts used in geography

Newly Industrialising Countries are among the fastest changing in the world. They are growing economically, changing socially and becoming powerful role models for other LEDCs. Rather than cover all the NICs, this section focuses on one NIC to illustrate some of the themes that can be developed. A city such as Singapore may also be used in any question that asks for a case study city.

Geography

Singapore is 648 km^2 (approximately 42 km × 23 km) – not much different in shape and size from the Isle of Wight – with some 60 smaller islands. It is 137 km north of the equator, so has a tropical climate, with a 26.7°C average temperature, 2350 mm rainfall annually and high humidity of an average of 84%. A map of Singapore is shown below:

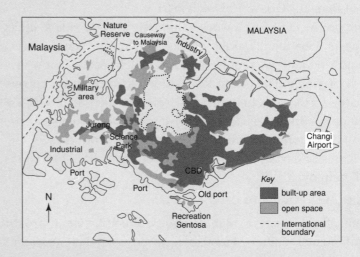

Indicators

	Life expectancy 1999	GDP/ cap 1998	GNP/ cap 1998	Pop/ tel	Pop/ doc	Infant mortality
UK	77	$20 212	$22 268	1.9	321	5/000
Singapore	77	$28 235	$31 900	2.0	667	3/000

Economic development

→ Pre 1965 – tied to Malaysia – entrepôt and regional trading centre
→ 1965 industrial estate launched at Jurong
→ Jurong Town Corporation 1968 attracted electronics firms
→ Development Bank of Singapore 1970
→ Attraction of major TNCs (electronics) to manufacture here
→ Development of Changi Airport as a major hub 1970–1980s
→ Expansion of port facilities, including oil refining
→ Regional HQ for service sector TNCs

→ Introduction of R&D rather than assembly due to labour shortages and high labour costs in 1990s

→ Creation of Science Park at Jurong next to National University 1990s

→ Production moved 'offshore' to Johore in Malaysia and Batam in Indonesia. Part of a 'growth triangle'

→ International Business Hub 2000 framework a service sector gateway to South East Asia

→ Further service sector developments such as Exhibition Centre

What factors enabled industry to develop?

→ Export incentives – 90% concession on export profits of foreign companies for five years – began 1967

→ Liberalised exchange controls – began 1978

→ Tax reductions on R&D work

→ Tax concessions on capital investment

→ Tax concessions for pioneer high-tech industries for ten years

→ Investment allowances

→ Singapore registered and owned shipping has tax concessions

→ Skills development fund to finance training

The figures below show the structure of the economy in 1960 and 1996.

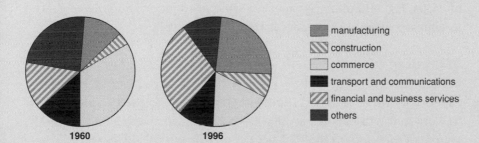

The jargon

R&D stands for 'research and development'.

Checkpoint 2

What is a *TNC*?

Check the net

www.singstat.gov.sg is the government web site that contains a wealth of data and links to other sites. Do remember that Singapore is a very controlled society and the web site will only contain what the government wants you to see.

Exam questions   answers: page 170

1   Describe the changes shown in the graphs above. Why has the economic structure changed? (10 mins)

2   Look at the climate graphs below and describe the characteristics of the climate. What are the effects of this climate regime on people and on the built environment? (8 mins)

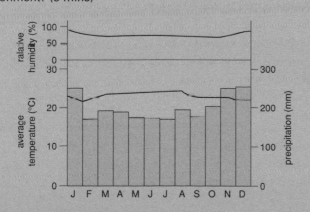

MEDCs

In describing the world as the home of people, it is impossible to look at every example of each phenomenon. In the study of population, economic and settlement geography it is appropriate to begin from an area within our own country as the most tangible set of examples. Because all parts of the UK are within a More Economically Developed Country (MEDC), AS and A-level Geography tends to draw many examples from England, Wales, Scotland and Northern Ireland. In this spread we assume that you are familiar with the British examples, and provided here are some examples from Italy that might prove useful.

Italy

Population 58 million, 67% urban, Rome 2.8 million, Milan 1.3 million, Turin 1.3 million, Naples 1.2 million. 1.3 children per woman, 8/000 deaths among under 5s.

Per capita GNP $19 300 and growing at about 1–2% per year.

Agriculture characterised by much poor, mountainous land (75% classified as hilly), especially in the spine of the country. Commercial agriculture mainly in the north (Lombardy Plain). Climatic determinism of crops, e.g. viticulture (grapes for wine production) and olives geared to climatic regime. Irrigation for rice in the Po valley.

Migration

In the twentieth century, Italy exported labour to Northern Europe, especially Switzerland and Germany. Many came to the UK in the 1950s to work in the brickfields around Bedford. Italians were also a major source of migrants to the USA between 1870 and the 1930s (85 000 in 1926) – especially migrants from the south and Sicily. Internal migration from south to north to work in cities in 1950s. By 1980s movement was more to central Italy as it industrialised.

	South to North	North to South
1956	116 000	39 000
1976	142 000	92 000
1986	104 000	72 000

Return migration from Europe and internally is a more recent phenomenon. Returners either go back to the home area to retire or with enough capital to develop a small business. Some only return to the industrial cities and take new jobs rather than return to their less wealthy area of origin. Returners are slightly older, reflecting the five-year contracts e.g. in Germany, or those returning ahead of retirement.

Tourism

→ Coastal tourism for Italians, e.g. much of the Adriatic coast and Lipari islands, and visitors especially from Northern Europe, along the Amalfi coast and Liguria

→ Alpine tourism in north together with ecotourism – on increase in mountains

→ World Heritage Sites – mainly cultural: Pisa, Venice, Florence

→ City-based short breaks: Rome, Florence, Venice, Sienna, Verona

→ Self-catering and second-home tourism: Tuscany (Chiantishire!), Umbria – purchase of abandoned farms

Regional divide

Italy demonstrates the concepts of **peripherality** and **distance from core**, a North–South divide. **Mezzogiorno** is the name for the south. The *Northern League* is a political separatist group borne out of objections to the amount of aid and government money going to the south. The maps below illustrate this division:

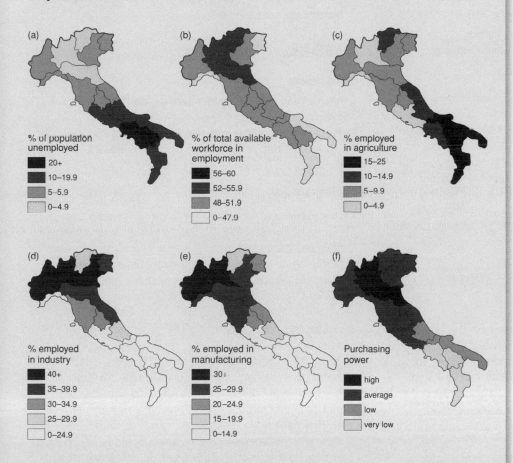

(a) % of population unemployed
- 20+
- 10–19.9
- 5–5.9
- 0–4.9

(b) % of total available workforce in employment
- 56–60
- 52–55.9
- 48–51.9
- 0–47.9

(c) % employed in agriculture
- 15–25
- 10–14.9
- 5–9.9
- 0–4.9

(d) % employed in industry
- 40+
- 35–39.9
- 30–34.9
- 25–29.9
- 0–24.9

(e) % employed in manufacturing
- 30+
- 25–29.9
- 20–24.9
- 15–19.9
- 0–14.9

(f) Purchasing power
- high
- average
- low
- very low

Exam questions

answers: page 170

1 Describe the main characteristics of Italy shown in maps a to f above. (5 mins)

2 What policies have been employed to redress the imbalances shown? (5 mins)

3 Some people refer to a 'third Italy' – the area south of the Po valley and stretching to Rome and Lazio. Why might such an area be increasing its economic strength? (5 mins)

Action point

Use some brochures from travel firms to locate different types of tourism in Italy (do not forget skiing brochures). How are the different types sold to the public?

Checkpoint 2

Do you know the meaning of *peripherality*?

Examiner's secrets

1 When dealing with regional disparities, whatever the country, do try to memorise a set of basic sketch maps that illustrate disparities.

2 Italy is not a developing country. You should not use the Mezzogiorno to illustrate the problems of a developing country – you will gain no marks.

Answers
Selecting countries

LEDCs

Checkpoints

1 *Ecotourism* is the concept of encouraging tourists to enjoy nature without spoiling it – a 'green' approach, e.g. a small group of tourists travelling with a local guide to observe wild animals rather than just driving through an area as a coach tour.

2 A *bustee* is an Indian name for a squatter settlement.

Exam questions

1 *Social* refers to problems connected with people, e.g. in India there are huge numbers living in squatter settlements as well as a caste system that divides society. There is the issue of population growth and control, as well as health issues, including AIDS. *Economic* refers to many of the problems that demography generates, e.g. coping with the housing, health, education and employment needs of such a large population. *Environmental* problems can be physical, e.g. failure of the monsoon rains, or human, e.g. pollution.

2 Choose an LEDC you have studied, or Kenya or India from this guide. Take each circle in turn, A–D, and then find its start point and explain the sequence, e.g. Population – the high birth rate in Kenya leads to large families of about five children. Because these families have low earnings due to lack of employment possibilities, living standards are low and only 23% of children attend secondary school; etc. You need to explain the key overlaps of the circles.

Singapore as an NIC

Checkpoints

1 GDP is Gross Domestic Product whereas GNP is Gross National Product. *GDP* measures the value of the goods and services produced by a country before providing for depreciation or capital consumption, whereas *GNP* is an evaluation of the overall performance of a country's economy including interest and profits earned abroad. *Infant mortality* – the average number of babies dying per 1000 live births.

2 A *TNC* is a *transnational corporation,* sometimes called a *multinational – these* are companies that usually have their headquarters in a developed country but have factories world-wide.

Exam questions

1 Main changes – decrease in commerce but a huge growth in manufacturing and financial and business services. Manufacturing has grown due to tax concessions on capital investment and pioneer high-tech industries. There are also export incentives. Financial and business services expanded to keep pace with the growth of manufacturing; Singapore also has an important function as an international business and exhibition centre.

2 Consistently high humidity both seasonally and diurnally. Consistently high rainfall – average 190 mm per month with November, December, January being slightly higher. Temperatures always around 24°C - a hot, humid climate. This climate is very tiring and all workplaces need effective air conditioning in order for people to be able to operate comfortably and efficiently.

MEDCs

Checkpoints

1 Milan, Turin and Genoa - known as the Industrial Triangle.

2 The *periphery* is the area that is on the edge of the main economic development (which is referred to as the *core*).

Exam questions

1 Unemployment highest in the south, especially Sicily. High employment rates in the north with unemployment gradually increasing on a N–S axis. The south has a low proportion of people eligible to work because there are many old people there – the young have moved in search of work. A high proportion of people in the south are employed in agriculture. In the north most people are employed in industry. North Italy generates most of the country's wealth.

2 Both the EU and Italian government have invested money in the south as part of a project called the *Casa per il Mezzogiorno* (The Southern Fund). This has invested money in infrastructure, irrigation schemes, industry and employment.

3 This region is between the prosperous Industrial Triangle and Rome, the capital. Both areas generate wealth and there is a good communication system between them. Investment and development is taking place along this axis between two important regions.

Successful study of geography at AS and A2 level requires you to be aware of the main techniques for studying the subject and for using your knowledge and skills to best effect in the examinations. Many students who have a good knowledge of the subject matter of geography do not perform as well as might be expected because they make errors in the way they interpret or answer questions, or in how well they demonstrate general skills, such as using examples, and geographical skills such as drawing sketch maps.

Being successful also requires you to show understanding of how the parts of geography fit together. Although you study your course in units or modules, these are only convenient ways of dividing the subject – they are not separate subjects, and there are many overlaps between them and common ideas that run through them. This section looks at some of the key aids to studying that will help you towards success.

Exam themes

→ Fieldwork data, enquiry and decision making

→ Using case studies and key skills

→ Illustrating your answers

→ Synoptic assessment

→ Projects, enquiries and investigations

→ Exam words/terminology

→ Effective revision notes

Resources section

Fieldwork data, enquiry and decision making

The jargon

Primary data is information collected from original sources, such as measurements of a stream's discharge or land use mapping in a city centre. It is data gathered in the field.

Secondary data is data obtained from published sources. The population census and meteorological records are examples of secondary data used by geographers.

Examiner's secrets

Remember that an example that you have found or seen is much better than the one in the textbook.

Watch out!

Avoid merely repeating the script of a video. The language is often a bit sensational. Always use your own words.

Field, primary and secondary data

Geography is about the world we live in and its complex interrelationships. To study that world we need to examine it first hand whenever possible. Therefore your AS and A-level studies will expect you to show that you have studied some of that world at first hand, through fieldwork.

Not all fieldwork has to be led by your tutor. Good tutors will take you into the field whenever possible and, if you are particularly fortunate, you may have spent a period of time away in a different environment. The examples that you see in the field, e.g. type of agriculture, an out-of-town retail park, an industrial estate, the impact of a landslide, should be used to support your written work in examinations. However, you should be able to use your own eyes, while journeying about your home area and while on holiday, to see other examples that you can use. These examples should be added to your own notes in the relevant place.

Primary data will come from fieldwork. Field measurements of stream discharge, maps of retailing in a city centre, or quadrant sampling of vegetation, are all examples of primary data collection.

On the whole, **secondary information** will have to suffice because you do not have the time or money to enable you to visit everywhere. One of the standard secondary sources will be the textbook that covers most of your specification, e.g. Michael Witherick, *Environment and People*, is a good general textbook. Other texts might relate to a specific module, e.g. S Frampton *et al.*, *Hazards*. Other good sources of information are journals such as *Geography Review* and *The Geographical. National Geographic* also has some relevant case studies such as the one on Hurricane Mitch in November 1999.

In this electronic age there is a whole range of secondary sources, some of which can almost be described as primary sources. The **Internet** is an excellent source if it is used carefully. For instance, the US Geological Service web site had details on the Turkish earthquakes shortly after they happened. Data on the Smoke Haze over South East Asia has been readily available on the Singapore Met Service site. The advantage of the web is that it can be up to date. Nevertheless be wary; some sites – especially those related to economic and social geography – may be biased because they are produced by pressure groups or companies with a specific marketing objective. Always see if you can find a site with an alternative view if you think that the opinions on one are too biased.

CD-ROMs are less up to date but they do form another valuable resource, especially for data sets. **Videos** are excellent visually.

Enquiry and decision making

Geography lends itself to enquiry-based learning. Enquiry has a specific route that you can follow, no matter what the topic.

→ *Describe* and *define* a phenomenon or issue – say **what** it is and **where** it can be found.
→ *Explain* the phenomenon or say **how** it came about, and *predict the outcomes* or suggest **what will happen if** something occurs.
→ Finally, one can *evaluate* and *prescribe* or say **how one should proceed** and how this solution or answer compares with alternative solutions.

A study of a stretch of coast might *describe* the coastal landforms of the mouths of Chichester Harbour and Langstone Harbour, *explain* how the spits were formed, *predict the outcomes* if people can roam all over the dunes at West Wittering, *evaluate* the measures to conserve the dunes and spits and, finally, *prescribe* the most satisfactory conservation solution on the basis of our knowledge of spits and dunes.

Decision making is the way in which many exam boards test your overall knowledge of your syllabus. It is a way by which the route to enquiry described above is used in an examination question to test your understanding of the subject. Your tutor will have given you practice throughout the year and the real test will probably come when you are issued with the examination materials, but not the questions, two weeks before you take the paper.

To be able to be a decision-making expert you need to:

→ understand and describe what the information shows
→ explain how and why the results that are given have come about
→ be able to rank solutions to the problem
→ explain why your rank is correct or why a solution might be more favoured than another.

Therefore you will notice that decision making is no different from the route to enquiry that you have probably used throughout your AS and A-level studies.

An exercise on global warming might:

→ ask you to describe data
→ explain the greenhouse effect
→ describe and explain the different attitudes people hold towards the greenhouse effect
→ evaluate some options to deal with the problem, and either say which solution you prefer, stating your reasons, or even develop your own strategy.

Action point

Find your own topic and see if you can fit your knowledge of it into this framework. It can help your revision.

Action point

Set up your enquiry on river floods, or a derelict site in the inner city, or migration into a city, or the impact of aid on a country.

Examiner's secrets

Decision making is *synoptic*, drawing your knowledge, understanding and skills together through the use of data from maps, photographs, tables, graphs and text to decide about a particular problem, challenge or issue.

A useful text on decision making is K Cowland, *Decision Making Techniques in Geography*, Hodder & Stoughton.

Action point

Could you follow this route using your own notes on the greenhouse effect?

Using case studies and key skills

Types of scale:
Macro-scale is the large scale such as the continent.
Meso-scale is about the size of a country.
Micro-scale is the smaller scale, the region.
Scale may also mean the proportion between length on a map and its true length on the ground, e.g. 1:25 000 means 1 cm on the map represents 250 metres on the ground.

Scale of study

Scale is an important concept in geography. The table below lists scale types and examples in some areas of geography at AS and A-levels.

Study focus	SMALL/LOCAL SCALE (S)	REGIONAL (R)	NATIONAL (N)	INTERNATIONAL (I)	GLOBAL (G)
Rivers	Small-scale catchment, e.g. River Meon Small-scale ecosystem, e.g. a wood or heathland in the New Forest	Large drainage basin, e.g. the Thames		Rivers with varied regimes, e.g. Mississippi/Rhine Impacts of rivers on people, e.g. floods Conflicts and consequences Management solutions	Hydrological cycle
Coasts	Landforms and processes on small section of coastline, e.g. Barton-on-Sea Sand dune and one other ecosystem, e.g. West Wittering Impacts of coastal change, e.g. Holderness	Extended coastline (E. USA) Coastal land uses and conflicts Management strategies	Coastal management strategies, e.g. The Netherlands	Contrasting examples of conflict, e.g. Mediterranean	Coastal environments Distribution of examples of coral reefs Sea-level changes
Rural	Changing villages e.g. in the Derbyshire Peaks Rural planning issues, e.g. in the Surrey greenbelt	Rural–urban continuum, e.g. the Vale of York Rural planning issues, e.g. in Hampshire Contrasts in rural areas, e.g. Cumbria Conflict and change in rural environments, e.g. a National Park		LEDC/MEDC contrasts, e.g. Ghana/France Two contrasting rural environments Future development strategies	Changing importance of rural/urban Global ruralisation
Urban	CBD changes, e.g. Birmingham Rural–urban fringe conflicts Managing urban environments, e.g. new towns Sustainable urban projects	Land use zoning in a large urban area, e.g. Bristol Managing urban environments, e.g. Bath	Cycle of urbanisation in one country, e.g. UK	LEDC/MEDC contrasts, e.g. London/Mumbai Contrasting world cities, e.g. Paris/Tokyo Quality of life Sustainable initiatives	Global urbanisation Global growth of megacities, e.g. Mexico City/Beijing
Population & economy	Population change at ward level, e.g. ward in a city Changing economic activity over time, e.g. South Wales	Regional population change, e.g. the North East Multi-ethnic conurbations, e.g. Toronto Inward investment in one region, e.g. M4 corridor	Policies of population and migration control, e.g. Scandinavia Role of migrant workers, e.g. Germany/UK NICs, e.g. Malaysia/Singapore Changing economic activity, e.g. UK/Germany	Demographic transition, e.g. Europe/Africa Over and under-population, e.g. Australia/Egypt Refugee problems Austria/Italy Links and disparities TNCs, e.g. Ford/Nestlé	Population dynamics Population growth Population/resource projections Migration, e.g. to New World Globalisation, e.g. car manufacturing Manufacturing or service industry study, e.g. banking Economic futures

Key skills in geography

The skills you should try to demonstrate as a geographer are:

→ Written English. All specifications now include marks for the quality of language. You must be able to organise, present information, ideas, descriptions and arguments clearly and logically. Therefore you must use correct grammar, punctuation and spelling.
→ The ability to interpret maps. The Ordnance Survey map symbols should be understood. Practise how to identify landforms and townscapes on maps.
→ Ability to interpret maps, diagrams and graphs.
→ The ability to draw sketch maps, diagrams and graphs to make points.
→ The ability to interpret tables of data.
→ The ability to interpret photographs whether they are normal *landscape* style, *oblique* or *vertical air photographs*. Also show that you can interpret satellite images either of weather systems or of landscapes (these use false colours).
→ The ability to process your own data gathered from primary sources.

The use of your knowledge is a skill. There are five types of knowledge that a geographer needs:

1 Geographical terminology, e.g. *desertification, counterurbanisation*
2 A range of locations, e.g. *a city, a stretch of coastline, the tundra*
3 Geographical processes, e.g. *migration, succession*
4 Geographical concepts, principles, theories and models, e.g. *suburbanisation, Smith's spatial margins theory of location, the core–periphery model*
5 Sources of geographical information, e.g. *censuses, sample surveys, hydrographs, diagrams and graphs.*

Applying your knowledge is the skill of understanding. You should be able to show:

→ the nature and interaction of processes – *how people respond to floods*
→ how places are distinct and interdependent – *why Paris is different and why its new towns are like those elsewhere*
→ how systems, patterns and places respond to change – *how coastlines respond to storms and people try to affect the changes*
→ how people perceive events and make decisions – *how people respond to hurricanes (either as politicians or as citizens)*
→ how data has limitations – *indicators of development.*

Illustrating your answers

Your grade in geography will depend in part on how well you can show that you know real examples of the ideas and issues that you are writing about. Geography is the study of the world around us, so you've got to show that you know case studies, and can both describe and illustrate them, for example with maps and diagrams.

Examples and case studies

Throughout your course in geography you should always include examples and case studies in your notes and in your written work.

→ An example is simply the name and a few brief details of a real place or situation, e.g. an example of a rapidly growing LEDC city is Mexico City, with a population of about 22.5 million and a growth rate of 5% per year.

→ A case study is a much more detailed example that you can use to discuss a range of issues relating to the topic. A case study of population growth in India would include at least the following ideas and information:
  → location of India
  → general economic characteristics: LEDC, GNP per capita about $350, rural economy, some industrial regions
  → basic demographic data (1941 – 319 million, 1998 – 989 million), birth rates (1941 – 45.9, 1998 – 28.0), death rates (1941 – 37.2, 1998 – 9.0), growth rate (about 1.9%)
  → population structure (e.g. population pyramids)
  → rural–urban contrasts (74% rural, 26% urban, but urban growth rate of 3% per annum)
  → regional contrasts (north – rapid growth, south – slower)
  → population policies.

You could record this information as notes, or on a map of India.

Building a range of examples

You may be asked to know examples in a range of different scales, from a range of countries at different levels of economic development, and for each of the topics you have studied. Check that you have the coverage by filling in a table like this one for each unit you have studied. This example has been filled in a for a 'population' unit:

	MEDC	LEDC
Local scale	Population in Hampshire	Population in Uttar Pradesh, India
National scale	Population change in UK	Population change in India
Continental scale	Migration to/in Western Europe	Migration in/from Africa
Global scale	World population growth + migration from South Asia	

Action point

Check your particular exam specification to find out what case studies it asks for in each unit. Check your own notes to be sure you have these case studies.

Action point

A good case study on India's population is by J Sanchez and A Marriott, 'Demographic Change in India', *Geofile* No 348, January 1999.

Action point

Draw up a table like the one here for each of the topics you have studied. If there are gaps, then check the specification to see if you need an example for that gap – if so, find one in a textbook and make notes on it.

Providing examples in exams

Whatever the form of assessment (traditional exam, timed essay, data response questions, etc.) you will be expected to give real examples in your answers. You may be asked to do this in various ways.

No instructions

Most questions do not ask you for examples, but you must still provide them. Illustrate each idea you present with a named example or a brief case study. Examples from your own fieldwork are a good idea.

With reference to specific examples/case studies

This is asking you to build your answer around one or two detailed case studies, e.g. a case study of managing visitor pressure at Tarn Hows in the English Lake District. Check the number of examples or case studies you are being asked for. Within the case study you will be expected to know the detail, i.e. names of places, production figures, dates of key events, names of organisations involved, etc.

Locating examples

Whether your examples are brief or detailed as in a case study, you must locate the example carefully. Say exactly where it is and, if possible, draw a simple sketch map. For a case study a sketch map is almost essential. Where you want to draw a map of the case study area **and** locate it in a region, country or continent, draw a small inset sketch map next to your main map to show where it is.

Important note!

Don't let your case studies and examples be a straitjacket for you.

→ Always pick out the important and relevant details to include. An exam question will ask you to use ideas from a case study. *Don't* just write down everything you know about the case study.
→ Mix and match the case studies and examples. Use some ideas from one case study together with some from another if necessary. Remember – because you learned a case study in a unit entitled 'urbanisation' doesn't mean you can't use it to answer a question on sustainable development if it is relevant and helpful.

Illustrating your examples

Sketches, maps, diagrams and charts are important in geography because:

→ they provide a visual image of data
→ they provide a useful summary of data
→ they demonstrate you can use key geographical skills.

In all exam mark schemes some of the marks are for illustrative material, so make sure you pick up those marks – even where the question doesn't specifically ask for illustrations.

Action point

Look at the questions in past exam papers or specimen papers. Against each one note down which examples and case studies you would use.

Action point

Get into the habit of summarising ideas or facts in a table rather than as text. It is quicker and easier than writing it all out. For example, the data on India supplied at the beginning of this spread could be presented as a table in an exam answer.

Action point

For *every* case study you have, practise a simple sketch map which you can draw in a minute or two in the exam room. Remember, it doesn't need to be a work of art – it's a quick sketch map!

Examiner's secrets

Many candidates simply throw away the marks which may be gained for good sketch maps or diagrams by not bothering to include them.

Synoptic assessment

Action point

Check your particular specification to find out how the synoptic assessment operates.

All A2 specifications have to have at least 20% of marks for 'synoptic assessment'. A synoptic assessment is designed to test how well you can:

1 understand the overall nature of geography as a subject
2 understand the links between the parts of geography
3 draw on that understanding to investigate new or different situations using your geographical understanding and skills.

This is normally tested through one or more of the following:

→ synoptic questions in an end-of-course exam
→ personal investigations/projects
→ decision-making exercises.

Synoptic exam questions

These are questions based on particular ideas or themes that draw together many aspects of geography. They are normally related to one of two broad fields:

→ people–environment interactions
→ sustainable development.

E.g. 'With reference to a major water resource development project explain how the exploitation of water resources can bring about physical, economic and social changes in the immediate environment.'

This question requires you to show knowledge from many parts of your course – water management and its links with hydrological processes (e.g. flooding), economic systems (e.g. water supply and farming) and social/political issues (e.g. the rights of local people *v.* industrial demands). You will also need to know a case study that shows these links, e.g. the Indus Valley Project in Pakistan.

Personal investigations

These require you to draw on your knowledge and skills to investigate an individual topic. This is described on pages 180–1.

Issues analysis

Issues analysis (formerly Decision Making Exercises) is an exercise in which you are asked to use your geographical knowledge, understanding and skills to analyse a particular problem, question or issue and to propose a solution or strategy for dealing with that issue. Examples of the issues you might be asked to consider are:

→ choosing between alternative routes development projects for an island
→ choosing a strategy for managing a popular visitor location
→ choosing the best site for a new power station
→ choosing between locations for a domestic waste disposal site
→ choosing a strategy for managing a river liable to flooding

Examiner's secrets

Check whether DMEs are included in your particular specification. If so:

→ How is it structured?
→ How are marks awarded for the different parts of it?

Action point

Pick a local issue that is currently in the news. Identify the interested parties and what their viewpoints might be.

A Issues Analysis can require you to deal with an issue in various ways:

→ Take on a role to examine the issue. This may be a role, for example, as a journalist, an academic, a planner, a local resident, a government official, or chair of an independent enquiry. In 'playing this role' you will be expected to show that you can understand an issue from a different person's perspective.

→ Study, analyse and interpret a range of information sources about the issue involved. These may be maps (at any scale), diagrams, tables of data, reports, summaries of the views of a range of interested parties, and technical information about the issue and the range of possible 'solutions'. This will test your ability to interpret geographical data and use appropriate geographical skills. Some of this information may be provided to you before the examination as a 'resource booklet' or 'advance information booklet', enabling you to analyse the data and undertake some research on the issue or location before you enter the exam room.

→ Show that you understand the nature of the issue by describing its causes and the social, economic and environmental factors that have an influence on the issue. In describing the nature of any issue these three headings (social, economic and environmental factors) provide a good framework for structuring your answer. To score high marks, you will be expected to show that you understand the links and interactions between these three groups of factors. A useful way to summarise these ideas is in a table:

	Social effects	Economic effects	Environmental effects
Positive impacts			
Negative impacts			

→ Show that you can identify the range of different viewpoints about the issue. You will be expected to be aware that people's views on an issue depend on:

→ their vested interests in the issue, i.e. what they will gain or lose from the issue

→ their own attitudes and values, which depend upon their background, experiences and personal 'view' of the world.

→ Critically review and analyse the existing proposals, using a range of techniques, such as cost-benefit analysis or environmental impact analysis. You may need to produce diagrams, tables or sketch maps.

→ Choose a preferred option, or identify a solution yourself.

→ Justify your choice with reference to the evidence.

You will also need to present your findings in a logical, well-argued way, using good spelling and grammar and making suitable use of graphs, tables and sketch maps.

Projects, enquiries and investigations

Projects are designed for you to show how you can pull together your geographical knowledge and skills, to investigate a specific topic or question. In particular they will test your skills of geographical enquiry (see pages 172–3).

Planning the project

Spend some time planning the project so that you know your deadlines, and what you must do to meet the exam requirements.

→ Work backwards from the date you must hand your work in.
→ Allow 2–3 weeks more than you think you need, to allow for unforeseen circumstances (illness, projects in other subjects, etc.)
→ Identify a data collection time over a holiday if possible.
→ Build in plenty of writing time – A2 projects may take 40–50 hours of writing time!

The figure below shows a timeline for an A2 project.

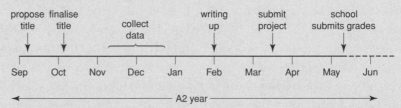

NB. Cycle is applicable *ONLY* to those specifications with an A2 project submitted as part of the June examination

Choosing a topic

If you are free to choose a topic, choose one that genuinely interests you, so that you are motivated to do your best work. Choose a topic that is:

1 One you found interesting in a unit you have studied, *or*
2 About an issue in the local news, e.g. plans for a new road, *or*
3 About an issue at a national or international scale (e.g. the decline in beef farming), but look at the issue in your own locality.

Key questions and hypotheses

Once you have chosen a general topic, you will need to identify a **key question** to ask, and/or a **hypothesis** to test. The key question may be the title of your project, e.g. 'What are the likely environmental impacts of proposed housing developments in Bursledon?' This will need to have several sub-questions following a route of geographical enquiry, e.g. *Why* is there pressure to build here? *Who* are the interested parties and what are their views? *What* alternatives have been put forward? *What should* be allowed to happen?

If you choose to test a hypothesis (e.g. 'Out-of-town shopping centres cause a decline in shopping services in the city centre'), the evidence should still be collected through a geographical enquiry route. The only difference is that at the end you will need to accept or reject the hypothesis.

The jargon

Projects are given many titles depending on exactly what they ask you to do, e.g. *personal enquiry, environmental investigation, personal investigation, personal investigative study.* Check what your specification calls it, and read the 'rules' for it.

Action point

Which two topics that you have studied have you found most interesting? For each, write down a question you could study for a project.

Action point

Read a copy of your local newspaper. Identify three current news topics that you could use for a project.

Action point

Investigating a local angle on a global problem provides a good project. Try to identify some local issues in your area arising from poverty in LEDCs (e.g. sales of 'fair trade' products in local shops).

Action point

For a project investigating issues about a proposed new housing estate on farmland, who might be the 'interested parties' you would need to contact?

Collecting background information

What is known already about this topic? Start by collecting information from a range of sources:

→ books, articles, newspaper reports, official reports, etc. from your geography department or library, or from your local library
→ the Internet
→ official bodies, e.g. the local council, charities, NGOs.

Collecting the data

Most projects require you to collect primary data, but you will also use secondary data as well. You will need to choose the most suitable research method to collect the data, and be able to justify your choice. You will also need to carry out the data collection in a reliable way. It is important not to collect too much data, or you will be overwhelmed by it – e.g. a sample of 50–100 questionnaires, or 10–15 interviews, or 10–20 soil pits/vegetation quadrants will be plenty.

Analysing the data

Data analysis has three stages:

1. Data presentation – use graphs, tables, maps, diagrams, overlays etc., but check you are using each method correctly. A good source book is D Richardson and P St John, *Methods of Presenting Fieldwork Data*, published by The Geographical Association.
2. Testing the data for relationships – either by using a statistical test (e.g. Spearman's Rank Test) or by examining it for apparent connections in the data. A good source is P St John and D Richardson, *Methods of Statistical Analysis of Fieldwork Data*, published by The Geographical Association.
3. Interpreting the data, to answer the key question or test the hypothesis.

Drawing conclusions

Your project must end with clear conclusions that:

→ summarise the main findings from your work
→ answer the key question, or accept/reject the hypothesis
→ critically review your study
→ make recommendations for managing the issue

Writing up/presentation

Think carefully about:

→ the report's structure – you will need sections on 'aims', 'background', 'research methods', 'results', 'conclusions'
→ quality of presentation – use your ICT skills to the full
→ spelling, punctuation and grammar i.e. Quality of Written Presentation

Action point

When contacting an organisation for information, be very precise about what you need, e.g. details of agricultural development projects in The Gambia, not 'anything on farming in LEDCs'. Always offer to pay for postage (or send a sae).

Action point

Find out the difference between *random sampling* and *systematic sampling*.

Examiner's secrets

A small-scale, highly focused project produces the best grades.

You will lose marks for poor spelling and grammar.

Gain marks by using a range of presentational methods, e.g. graphs, tables, photographs.

Exam words/terminology

Practise with sample exam papers by underlining the key command words in each question.

Think how you would answer this question: 'Discuss the issues involved in considering a planning application for a new opencast coal site'.

Examiner's secrets

Geographical enquiry questions provide a useful checklist for what you should be trying to answer in an exam question.
(Describing) What? Where? When? Who? What effect?
(Explaining) How? Why?
(Discussing) above + What will? How should?

Action point

Ensure that the case study you choose to answer a question is relevant to the topic – and only include those aspects of the case study that help answer the question. For example, if a question asks about social issues arising from nuclear power, don't write too much about the environmental impacts of your case study of Three Mile Island.

To make sure that your grade is as high as possible you must answer exam questions as carefully as possible. In particular, you must answer the question that *has* been asked, not the question that you *wish* had been asked. Understanding the precise meaning of the *command words* in exam questions is vital, as is checking that you are using the right type, scale and location of *examples*, covering the right geographical *concepts/ideas*, and that you are demonstrating the right geographical *skills*.

Key command words

Command words are the words that tell you what to do. The principal command words used are:

Describe . . .

E.g. 'Describe the main features of a hurricane.'

This is asking you to show that you know what is occurring in relation to a particular topic. You will need to show that you know **where** in general this occurs (say *exactly* where hurricanes occur), **what happens** (describe the main weather features of the hurricane), and the main **effects** of what is happening (e.g. wind damage, floods, disruption, deaths).

Explain . . .

E.g. 'Explain the causes and consequences of rural–urban migration in LEDCs.'

This is asking you to show that you know *how* and *why* something occurs. You will need to *explain* enough to show that you understand what is going on, but most of your answer will be the reasons for and causes of whatever you are being asked about. In the example question, start by saying briefly what rural–urban migration is, where it occurs in LEDCs, and some of the consequences. Then take each cause in turn and show how the process works (e.g. push factors such as rural population growth, pull factors such as employment), then discuss each consequence (e.g. growth of shanty towns) and the reasons for it.

Discuss . . .

E.g. 'Discuss the consequences of the development of 'out-of-town' shopping centres.'

If you are asked to discuss a topic you will need to show a wide range of knowledge and understanding. It includes both *describing* and *explaining* the topic under discussion, but also drawing out some conclusions, e.g. do the advantages outweigh the disadvantages (or vice versa), what are likely to be future impacts, changes or developments? What could or should be done to deal with the issues raised?

Compare/Contrast . . .

E.g. 'Compare and contrast the flood hydrographs of an urban and a rural catchment.'

This command is asking you to look at two (or more) different situations and show how they are both similar and different. You will need to describe the similarities and differences but also be able to explain them, and show the consequences of any significant differences. In this example question you would describe differences in the hydrographs (e.g. shape, peak flows, lag times, flooding risk) and then explain them (e.g. surface permeability, vegetation cover, etc.).

Critically evaluate . . .

E.g. 'Critically evaluate the range of solutions available for reducing soil erosion caused by agricultural practices.'

This is asking you both to describe and to explain a topic and also to draw some conclusions on the basis of evidence. You will need, for this example, to discuss each solution available and indicate its pros and cons, and decide how likely it is to be a success, presenting evidence for your decision. Remember that 'being critical' involves looking at both positive *and* negative things, not just the negative ones.

Other key words

Using the right examples

Check the question to be sure that you are using the right examples. Is the question asking for a case study or examples? Does it want one or two? Should those examples be drawn from particular locations, e.g. 'with reference to a city in an LEDC country . . .' is very clear about what it is asking. Write about Nairobi, or São Paulo, but not about London or Tokyo.

Using the right key concepts or ideas

Many questions are asking you to show that you understand particular concepts, and you need to be sure that these are the ones you cover in enough detail, e.g. 'with reference to a range of urban structure models . . .' Is asking you to look at several models you know and could critically review (e.g. sector models, LEDC urban structure models, etc.) – do not write too much about urban growth except about how it helps explain urban structure.

Using the right geographical skills

Questions ask you to draw on particular skills, either by asking you to draw a sketch map or a diagram or by drawing on examples from your fieldwork. Be sure to include this if it is asked for, as there will be marks awarded specifically for this.

Effective revision notes

Test yourself

For the topic you are currently studying in geography, write down three key general ideas and the location of a case study you could use to illustrate each idea.

Action point

How could you abbreviate the following in your notes:

➜ 'because'
➜ 'the government'
➜ 'increases'?

The notes that you have taken throughout your study programme are the key to your preparation for the examination. They not only provide the information and material that you should learn but also provide a way of organising your knowledge to make it easier to understand and remember.

What should be in revision notes?

By the end of a course unit your notes should contain the following:

➜ ideas (concepts) and information (knowledge/content) from your work in class and private study
➜ details of case studies and examples to illustrate the ideas and principles of the topic studied
➜ your answers to any test questions or exercises you have done.

Making notes

Remember that revision notes are notes, not full, grammatically correct essays. They need to be fully understandable by *you*, but not necessarily by anyone else. They should:

➜ use abbreviations and your own shorthand way of writing things, e.g. write 'pptn' rather than precipitation, 'ag' instead of agriculture
➜ use symbols/signs to show ideas, e.g. use → instead of 'causes'
➜ be organised in an easy way to read, e.g. use bullet points or lists
➜ make your notes visual, e.g. use spider diagrams or concept maps to show how ideas are linked, or use annotated maps or diagrams to summarise ideas. The figure below shows such a concept map.

Action point

Watch a TV documentary with a friend. Both take notes, then compare notes to see how each of you can learn from the other's style. You could try this with your whole class.

Action point

List the geography units in your course. Against each one list the case studies you have looked at during your course.

Making notes from lectures (including videos)

When you are listening to a lecture or talk, or watching a video:

➜ do *not* attempt to write down every word as if by dictation
➜ write down key words or facts as they occur
➜ make sure you don't miss the next idea by being too busy writing
➜ at the end of the lecture/lesson, take 2 or 3 minutes to go back through the notes and underline or highlight the main ideas (*not* the main facts).

Making notes from written resources

The secret of effective notes from reading resources is in the way you do the reading.

→ Never start taking notes until you've finished reading the resource.
→ Start by reading the introductory paragraph and the concluding paragraph, to give you a sense of what the item is about (and whether it is of any use to you!).
→ Then read the whole item carefully.
→ Summarise the resource in the following way:
  → Note the title, author, date and source of the resource.
  → Write a one-sentence summary of the resource.
  → Summarise the main ideas, using bullets, spider diagram, annotated sketch map or whatever is best for you.
  → For each idea, write down one specific example.
  → Keep your notes to no more than 20% of the length of the original, i.e. summarise a five-page article in no more than one side of notes.

Revising from notes

Remember that the aim of revision is not to clog your mind with facts but to ensure that you understand the concepts and ideas you have covered, with enough detailed knowledge to illustrate and exemplify those ideas. In revising from your notes:

→ start by reading through all your notes on a topic or unit to re-familiarise yourself with what you have covered
→ identify the key ideas you should understand in the topic by reading the relevant section of the specification that you are studying
→ for each key idea in the specification, produce a spider diagram or list of bullet points of the relevant things in your notes, e.g. for LEDC urbanisation you may have notes from two lessons, notes from a video on São Paulo, notes on São Paulo from a *Times* article, notes on a *Geography Review* article on Nairobi and Cape Town
→ then revise each key idea by reading through the relevant sections of your notes – you may need to do this 5–10 times
→ after each reading of your notes, try to recall as many of the key points as possible by writing the ideas down on paper
→ to revise the skills recorded in your notes, try practising each one, e.g. practise a chi-squared test, using some data in a textbook
→ look at past papers and questions to identify whether your understanding from your notes would enable you to answer them – if not, go back to a textbook to take some more notes.

For the final stages of revision, you may find it useful to summarise the main points and details of case studies for each key idea from the specification onto a single 'card index' card.

Action points

Organise bullet point lists so that the first letters spell a word or are in alphabetical order. This makes memory easier, e.g. river erosion occurs through CASH:

→ **C**orrasion
→ **A**ttrition
→ **S**olution
→ **H**ydraulic action.

Draw a spider diagram to show the causes and consequences of global warming.

Examiner's secrets

Some case studies link to several topics in your course, e.g. agriculture in Kenya may relate to agriculture, development, environmental hazards, tropical savannas, etc. Add a note to your notes in every relevant place to look at such case studies.

"Knowledge should be a light to the mind not a burden on the memory."

Anon

Examiner's secrets

Notes are good for revision, but your written answers must be in full sentences and correct English. Only use notes as annotations to diagrams, etc.

Index